服饰

Ancient Costumes and Accessories

沈周 ◎ 编著

图书在版编目(CIP)数据

服饰 / 沈周编著. -- 合肥：黄山书社, 2015.11
（印象中国. 纸上博物馆）
ISBN 978-7-5461-4209-8

Ⅰ.①服… Ⅱ.①沈… Ⅲ.①服饰文化—中国—古代
Ⅳ.①TS941.742.2

中国版本图书馆CIP数据核字(2015)第275829号

服饰　　　　　　　　　　　　　　　　　　　　　　　　沈周 编著
FU SHI

责任编辑	程　景
责任印制	戚　帅
图文编辑	王　新
装帧设计	李　晶　王萌萌
出版发行	时代出版传媒股份有限公司（http://www.press-mart.com）
	黄山书社（http://www.hspress.cn）
地址邮编	安徽省合肥市蜀山区翡翠路1118号出版传媒广场7层　230071
印　　刷	安徽新华印刷股份有限公司
版　　次	2016年8月第1版
印　　次	2016年8月第1次印刷
开　　本	720mm×1000mm　1/16
字　　数	143千
印　　张	11.25
书　　号	ISBN 978-7-5461-4209-8
定　　价	39.00元

服务热线　0551-63533706　　　　　　版权所有　侵权必究
销售热线　0551-63533761　　　　　　凡本社图书出现印装质量问题，
　　　　　　　　　　　　　　　　　　请与印制科联系。
官方直营书店（http://hsssbook.taobao.com）　　联系电话　0551-63533725

前 言 Preface

　　服饰是人类生活的要素，也是一种文化载体。在中华五千年的文明史中，服饰承载着厚重的传统文化内涵，它不仅可以驱寒蔽体，而且满足了古人的审美情趣，反映了古代的民俗风情和社会制度，记录了历史和社会生活，体现了人们的思想观念和民族精神，彰显了社会等级制度。

　　中国素有"衣冠之邦"的美称，服饰历史源远流长。从商周时期的冠冕之制、胡服骑射，再到唐代的开放女装、清代的旗袍，中国服饰以其鲜明的特色

Costumes have been conveying the connotation of traditional Chinese culture for five thousand years of Chinese civilization. Acting as an element of human life, costumes not only protect people from being cold, but also meet the ancient people's aesthetic taste. What's more, serving as a carrier of Chinese culture, they reflect the ancient folk customs and social systems, record the history, social life and express people's mentality, national spirit and highlight the social hierarchy as well.

　　Because of the long standing and well-established history of costumes, China is known as the "Kingdom of Dresses". Looking back at the costumes history of China, from its Crown institution of Shang and Zhou Dynasties (1600B.C.-221B.C.), wearing *Hu* dress shooting arrow from horseback, then to the open style of women's

为世界所瞩目。中国古代的先民们不但创造出精美绝伦的服饰，更创造了丰富多彩的服饰文化，为历史增添了绚丽的色彩。

请让我们一同走进异彩纷呈的中国古代服饰世界，体味龙袍凤冠的华贵与厚重，感叹霓裳羽衣的绚丽多姿，品鉴那绝代芳华。

clothing in Tang Dynasty (618-907), to Chipao of Qing Dynasty (1644-1911), Chinese ancient ancestors not only created numerous exquisite costumes, but colorful costume culture which added bright color to Chinese history.

Now let us walk into the colorful world presented by ancient Chinese costumes. Wandering in this world, you can savor the luxurious Dragon robe, Phoenix coronet and be amazed by the gorgeous costumes and accessories.

目录 Contents

霓裳古国
Costumes of Ancient Chinese 001

服饰的起源
Origin of Costumes 002

服饰与礼制
Costumes and Social Etiquette 006

龙袍与凤冠
Dragon Robe and Phoenix Coronet 013

冕服
Ceremonial Robe 014

龙袍
Dragon Robe 021

凤冠
Phoenix Coronet 027

朝服与官服
Court Costume and Official Costume 035

朝服
Court Costume 036

官服
Official Costume 040

便服
Casual Wear 057

深衣
Shenyi (Garment) 058

袍
Gown-Style Dress 065

衫
Shan (Long Gown) 069

襦
Ru (Short Jacket) 076

袄
Ao (Short Jacket) 081

褂
Gua (Jacket) 084

裳
Shang (Dress on the Lower Part of the Body) 087

裙
Skirt 096

裤
Pants 109

背心
Sleeveless Jacket 114

内衣
Underwear 117

巾、帽
Scarf and Hat 121

鞋、袜
Shoes and Socks 133

胡服与戎装
Hu Dress and Martial Attire 143

"胡服骑射"
Wearing *Hu* Dress and Shooting on
Horseback 144

甲胄
Armor 148

婚服与丧服
Wedding Dress and Mourning Dress 157

婚服
Wedding Dress 158

丧服
Mourning Dress 164

霓裳古国
Costumes of Ancient Chinese

  中国的服饰历史源远流长，早在原始社会就已经有了能够蔽体的最早的衣服。随着社会的发展、文明的进步，人们对于服饰的要求不仅仅是遮身暖体，而是越来越追求美感。自此，服饰文化蓬勃发展起来，人们将生活习俗、审美情趣、色彩纹样以及文化和宗教都融于服饰之中，构成了特色鲜明、博大精深的中国服饰文化。

China boasts a long standing history of costumes dating back to the earliest primitive society. With the advancement of China's social culture and civilization, people demanded clothes to be more than a covering of the body to keep warm. People increasingly focused on the pursuit of beauty, which led to the development of Chinese costumes. People integrated customs, beauty, color patterns, culture and religion into the costumes, which led to formation of the distinctive and profound Chinese clothing culture.

> 服饰的起源

远古时期，人类最早是用兽皮、树叶、茅草、鸟羽等来遮身暖体，这就是服装的雏形。直到距今约一万年前，人类进入了新石器时代，中国的先民们逐渐掌握了纺织技术，始拥有了真正的"衣服"，从茹毛饮血的混沌时代迈出了走向文明的脚步。

> Origin of Costumes

From time immemorial, human beings used animal hides, leaves, grass and feathers to keep warm. These were the prototypes for the original clothes. Dating back to 10,000 B.C., people entered into the Neolithic Age. Chinese ancestors gradually acquired textile technology ushering in a civilized society where people had a real sense of "clothes".

Linen, ko-hemp, silk, and wool were the main clothing materials at that time. In the ruins of 6000 years old *Yangshao* culture, ten warps and ten welts woven coarse sacking was found on each square

- 原始人生活场景

原始人最初只是用兽皮、树叶等当衣服围住下体。

Scenes of primitive life

The primitive men use animal hides and leaves to shield the lower part of the body which is the original clothes for the primitive men.

- 浙江湖州出土的新石器时代的丝线、丝绳和麻布残片

Unearthed thread, string and linen fragment of the Neolithic age in Huzhou, Zhejiang Province.

- 余姚河姆渡遗址出土的陶纺轮

纺轮是早期的纺织工具，以陶制、石制为主。

Unearthed pottery spinning wheels in Hemudu, Yuyao

Spinning wheel is a tool for spinning and weaving in the early times, mostly made of ceramic, and stone-base.

　　那时的服装材料主要为麻布、葛布、丝绸、毛织品等。在距今6000年前的仰韶文化遗址中，曾发现了每平方厘米经纬各有10根的粗麻布印痕；在4000年前的良渚文化遗址中，发现过每平方厘米经纬各有20至30根的织布。

　　此外，考古人员还在良渚文化遗址中发现了丝绸残片，虽然这些丝绸残片已有轻微的碳化现象，但仍保持着良好的弹性，密度为每平方厘米经纬各48根丝线。这说明，当时的中国人就已经掌握了养蚕、取丝、织绸的技术。纺织技术的发

centimeter. In ruins of 4000 years old *Liangzhu* culture, twenty to thirty warps and wefts woven cloth was found on each square centimeter.

In addition, archaeologists found fragments of silk in ruins of *Liangzhu* culture. Although these silk relics were slightly carbonized, they still retained good elasticity. Its density reached 48 silk threads warps and welts on each square centimeter. It showed that Chinese people had mastered the skill of raising silkworms, processing the silk and silk weaving technology. The invention of textile technology opened a new page for

明，为后来中国灿烂的服饰文化开启了崭新的一页。

原始人类不仅发明、制作了服装，还添加了很多附属饰件进行美化。如山顶洞人就已懂得用穿孔的兽牙和小石子作为装饰品；仰韶文化遗址中也发现了戴在颈间的骨珠、动物牙齿、蚌珠蚌环，以及骨笄、骨簪等头饰。

the development of the splendid Chinese clothing culture.

Primitive people invented and made clothes. They even added a number of ornamental accessories to beautify the clothes. For example, the Upper Cave Man already knew how to use bored animal teeth and stones as ornaments. Bone beads, animal teeth, mussel shells, hairpins and other hair accessories were discovered in the ruins of Yangshao culture.

- 邯郸出土的贝饰
Unearthed shell accessories in Handan.

- 邯郸出土的蚌佩
Mussel pendants are unearthed in Handan.

- 陕西临潼出土的兽牙装饰品
Animal's teeth ornaments are unearthed in Lintong, Shaanxi Province.

第一支缝衣针

大约在18000年以前,生活在北京周口店龙骨山山洞里的原始人就已经懂得采集、狩猎、使用石器和人工取火了。考古人员还从其生活遗址中发现了一支骨针。这支骨针经过磨制,长达8.2厘米,针身光滑,直径3.1至3.3毫米,且针尖十分锐利,针孔直径只有1毫米。

骨针的发现,说明那时的中国人已经懂得缝制衣服,中华服饰史自此发端。只不过缝制的材料是兽皮,而缝线可能是劈开动物的韧带得到的丝筋。

The First Sewing Needle

Around 18000 years ago, the primitive people living at the cave of the Dragon Bone Hill, Zhoukoudian, Beijing had known how to gather, hunt and make fire with stones. Archaeologists also discovered a bone needle among the ruins. It was smoothly grounded, 8.2 cm long and had dia. of 3.1mm to 3.3 mm. The needlepoint was very sharp and produced a 1mm diameter pinhole.

The discovery of the bone needle indicated that early Chinese people started to sew the clothes. They used animal's split ligaments as thread to sew the hides. Therefore, the history of Chinese costumes originated ever since then.

- 原始人使用的骨针

Bone needles are used by the primitive.

> 服饰与礼制

中国自古崇尚礼制，在服饰文化中也有明确的体现。《易·系辞下》中说道："黄帝、尧、舜垂衣裳而天下治，盖取诸乾坤。"这句话的意思是黄帝、尧、舜用衣裳来区分管理者和普通百姓，治理国家。可见从这时起，服饰已经明确地与治理国家的道理联系在一起了。

上身有衣，下身有裳是中国最早衣裳制度的基本形式。古人认为，天在没亮的时候是玄色（即深蓝，近于黑色），因此将上衣规定为玄色；地是黄色，因此把下裳规定为黄色，中国几千年来的"上衣下裳"制度便由此形成，直到今天人们仍把服装称为"衣裳"。

奴隶社会建立后，中国的礼乐衣冠体系逐步形成，上至天子，下

> Costumes and Social Etiquette

China has been advocating social etiquette since ancient times, which is shown in the culture of costumes. As mentioned in the ancient book of *Yi-Xicixia*, the "Yellow Emperor, Yao and Shun distinguish the governors and common people from what they wear. They govern the nation by different levels of dress." It's safe to say that costumes have been associated with governance of the country since then.

The basic form of the earliest costumes in China was composed of a Yi (coat) on the upper part of the body and a *Shang* (skirt) on the lower part of the body. In ancient times, people chose the color of dawn as *Xuan* (dark blue, very close to black), as the color of this coat. The color of dirt was yellow brown and became the color of skirts. *Shang Yi Xia*

● 黄帝像
Portrait of the Yellow Emperor

至庶民，无论贵贱尊卑都应穿着相应的服饰。服饰作为一种文化现象始终被礼制所约束。尤其到了封建社会，等级制度在服饰上更有着极其显著的反映。它与礼制紧密结合，规范了不同阶层穿衣戴帽的制度，对服装的质料、颜色、图案纹样等都有详尽的规定，以区分君臣士庶的身份和地位。即使是在相对开放的唐代，服装样式虽千变万化，但也不能逾矩。一代女皇武则天就曾以绣袍赐予百官，这些绣袍是以鸟兽纹样为主，而且装饰在前襟后背等部位，体现了中国服饰的礼制文化特点。

Shang literally translated as "upper coat, lower skirt" and became the standard clothing. The name *Yishang* is still used today.

China's ritual dress system gradually took shape after the establishment of a slave society. From the emperor down to ordinary people, regardless of hierarchy, everyone shall wear appropriate clothing. Costume, as cultural phenomena, has always been bound by social institutions. Costumes closely integrated with China's etiquette especially in the feudal society, which were standardized in great detail so as to distinguish the status of ruling class from common people. Even in the relatively open Tang Dynasty, when different styles of dress sprang up, regulations on costumes were still strictly enforced. Empress Wu Zetian has ever conferred embroidered gowns on her officials. The gowns mainly had the patterns of embroidered birds and animals on the front piece and back, which reflected the feature of social institutions.

中国古代服饰演变图表
Chart Of China's Ancient Costumes Evolution

时代 Ages	服饰特征 Features of the Costumes	示意图 Illustration
原始社会（前30世纪—前21世纪）Primitive society (30th century B.C. - 21th century B.C.)	在原始社会，人们所穿并非真正意义上的服装，而是用树叶、茅草、鸟羽、兽皮等蔽体。从挖掘出的山顶洞人骨针来看，当时的人类已开始懂得简单的缝纫了。 What the primitive people wore was not the true sense of clothes as we called today. Leaves, grass, feather and animal hide were used to wrap the body. Based on our discovery of the bone-made needle made by Upper Cave Man, We could see they had known how to sew at that time. 后来，古人发现了麻、葛等，用以编织成衣物，原始的纺织业兴起。原始社会晚期，古人又学会了养蚕纺丝，制作出了丝绸。据《易·系辞下》记载，黄帝、尧、舜统治时期，中国的服饰制度就已初步建立，服饰也成为礼制的载体，上衣下裳为中国几千年来一直沿用的礼服形式。 With time passing by, the ancient people discovered hemp. Using this material, they wove dress and the primitive textile industry rose ever since then. The ancient people started to raise silkworm to spin and make silk in the late primitive society. As recorded in *Yi-Xicixia*, during the reign of Yellow Emperor, Yao and Shun, the system of China's costumes and accessories has been initially established. It has become a vehicle of social institution and continued to use in following several thousand years of Chinese history.	 • 原始社会的服装 Clothes of the primitive society

夏朝君王启的画像
Portrait of Monarch Qi of Xia Dynasty

汉代步兵俑
Infantry figure of Han Dynasty

| 夏商周
（前21世纪–前256）
Xia, Shang and Zhou Dynasties
(21th century B.C.-256B.C.) | 夏商是中国服饰等级制度愈加分明的时期，确立了"六冕""六衣"等服饰制度，说明服饰已成为权力和等级的象征。
In Xia and Shang Dynasties, a rigid hierarchy of China's costumes became all the more distinct. The hierarchical system of "six-crown and six-gown" was set up. It demonstrated that costume has been an attribute of power and hierarchy.

西周时期，一种名叫"深衣"的连体式长衣流行开来。深衣不分尊卑，无论男女都能穿着。
A kind of long garment named *Shenyi* became popular with no distinction of hierarchy and it's allowed to be worn by either men or women.

战国时期，赵武灵王进行了中国服饰的一场大变革，即"胡服骑射"，人们开始穿合身、轻便的服装。
During the Warring States period, King Wuling of Zhao State performed a profound reform "wearing *Hu* dress and shooting on horseback" that is, learning to ride and archery from the western ethnic minorities and starting to wear their costumes. Following this trend, people started to wear formfitting and light clothes. |

| 秦汉
（前221–公元220）
Qin and Han Dynasties
(221B.C.-220A.D.) | 秦始皇统一中国后，服饰制度也随之统一。
As First Emperor of Qin (Qin Shihuang) reunited the states, costumes had unified standard.

到了汉代，男子以袍服为贵，并作为礼服。女子则以深衣为贵，普通女子上身穿短襦，下穿长裙。
By Han Dynasty, men wearing gown-style dress which was taken as formal and noble attire, while women wearing *Shenyi* which was taken as noble attire. Ordinary women wore short coat and long skirt.

汉代还严格了舆服制度，即用首冠和佩绶区分身份等级。
Yufu institution was set up in Han Dynasty, that is, distinguishing ranks by crown and *Peishou* (ribbons hanging jade ornaments). |

魏晋南北朝 (220–589)
Wei, Jin, Northern and Southern Dynasties (220–589)

魏晋时期流行宽衣博带的服饰风尚，男子大都穿大袖宽衫，女子则上俭下丰，以"杂裾垂髾"为代表服饰。

Loose style costume prevailed in Wei and Jin Dynasties. Men wore loose clothes with large sleeves. Women's typical costume was *Za Ju Chui Shao* dress, which featured wide top and tapering bottom.

南北朝时期民族大融合，出现了"裲裆""袴褶"等服饰，男女的发式也更为讲究。

Northern and Southern Dynasties witnessed a large-scale merging of ethnic groups. Such costumes as Liangdang and Kuzhe appeared. People particularly cared about their hairdo.

- 杂裾垂髾服
  Za Ju Chui Shao dress

隋唐 (581–907)
Sui and Tang Dynasties (581–907)

隋唐是中国经济繁荣、政治稳定的时期，服饰自然也雍容华贵。尤其是女子的服装，款式繁多，色彩缤纷，非常浪漫开放。唐代女子通常穿短襦长裙，裙腰系得很高，开元年间还出现了袒胸装，甚至流行过"女穿男服"。

China underwent economic prosperity and political stability during Sui and Tang Dynasties. Costumes at that period of time were dignified and graceful, especially for women's costumes, they were rich in color and style. Women in Tang Dynasty usually wore short coat and long skirt with high-tied waist. Topless dress appeared in Kaiyuan Years at Emperor Xuanzong. "Women wearing men's clothes" was even popular at that time.

- 唐代仕女
  The maidens painting in Tang Dynasty

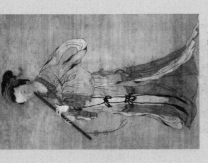

- 宋代贵族女子常服像
  Portrait of noble women in informal dress

- 元代官员像
  Portrait of the official in Yuan Dynasty

宋代
（960-1279）
Song Dynasty
(960-1279)

宋代的服饰偏重于清新自然，男子大多穿圆领大袖衫，女子上衣有襦、袄、衫、褙子、半臂等形制，下裳则以裙为主。
Costumes of Song Dynasty were shifted to be simple and natural. Men mostly wore clothes with loose sleeves and round collar. Women's clothes for upper part of the body included short coat, Ao, underlinen, Beizi and half-sleeve garments. Clothes for lower part of the body were skirt.

此外，宋代流行女子缠足，鞋成为女子的主要足服。
In addition, women in Song Dynasty bound feet, thus shoes became one of the major wears.

辽、夏、金、元
（907-1368）
Liao, Xia, Jin and Yuan Dynasties
(907-1368)

辽、夏、金、元是少数民族建立的政权，因此在服饰上，少数民族与汉族融合的现象日益频繁。这一时期的服装以长袍为主，一般为窄袖、圆领、长筒靴。
Ethnic minorities established the regime in Liao, Xia, Jin and Yuan Dynasties. There's costume integration between Han nationality and ethnic minorities. Gown-style dress with narrow sleeves and round collar to match boots became the main costumes at that time.

元代的服饰制度还对帝王、百官的服色作了统一规定，甚至区分了蒙古族与汉族的官服。
Unified regulations were made for the color of the emperor's clothing and officials' clothing in Yuan Dynasty. Mongolian official dress and Han official dress were distinguished.

明代
（1368—1644）
Ming Dynasty
(1368-1644)

明代服饰又恢复汉族传统，并重视礼制，严格规定了皇帝、百官的服饰品级，礼服与常服区分明显。

Costumes of Han were resumed in Ming Dynasty. Strict rules for costumes were set up. Social institutions were stressed. Ranks for emperor's and officials' costumes were formulated. Formal attire and casual wear were differentiated.

明代官员像
Portrait of the official in Ming Dynasty

清代
（1644—1911）
Qing Dynasty
(1644-1911)

清代是满族建立的政权，因而服饰特点鲜明，并建立了一套完整而繁缛的服饰制度。男子以长袍马褂为主，满族女子穿旗装，汉族女子则沿袭上衣下裙的服式。

Qing Dynasty was established by Manchu. At that time, distinctive clothing feature and heavy-coated style were formed. Men mainly wore long robe and ridding jacket, Manchu women with Manchu-style dress. Clothes for women of Han nationality still retained the style of upper coat and lower skirt.

清代满族女子的旗装
Manchu women's dress in Qing Dynasty

龙袍与凤冠
Dragon Robe and Phoenix Coronet

　　在中国古代文化中,服饰具有十分特殊的地位,它不仅用于穿着和美化,还用于区分社会等级,维护皇权,具有政治意义。在中国历代学者撰写的正史中,几乎每一部都有关于皇朝盛典服饰的记载,足见其重要的地位。而在这些具有政治色彩的服饰中,冕服、龙袍、凤冠等可谓代表了社会的最高等级,象征着至高无上的皇权。

In ancient Chinese culture, costume has its special significance. It's used to keep people warm and beautiful. What's more, costume was endowed with a political meaning so as to distinguish social class and maintain imperial power. In biographical history books written by ancient Chinese scholars, royal costume for almost every grand ceremony was recorded. Such these political related costumes as ceremonial robe, dragon robe and phoenix coronet stood for the highest rank of society and represented the supreme imperial power.

> 冕服

　　冕服是中国古代皇帝、公卿、士大夫的祭服，即祭祀活动的礼服。从西周开始，中国的衣冠服饰制度逐渐完善，专设有"司服"的官职，上至天子，下至群臣，参加各种礼仪活动时所穿的服饰皆有定制，如祭祀有祭服，朝会有朝服，处理公务有公服，服丧有丧服等。

　　按照西周的礼仪规定，天子、公卿、士大夫参加祭祀，必须身着冕服，头戴冕冠。冕服均为玄衣纁裳，以图案和冕旒的数目不同来区分身份等级。如帝王的冕服，上面绣有十二章纹，冕旒为十二旒。帝王在最隆重的场合穿绘绣十二章纹的冕服，其他场合减少章纹、冕旒的数目。公卿、侯伯随帝王参加祭祀时，冕服上的章纹和冕旒数量会

> Ceremonial Robe

Ceremonial robe was formal attire for Chinese emperor, grand councilor or scholar-bureaucrat to attend ceremony of offering sacrifices to gods and ancestors. Beginning from Western Zhou Dynasty (1046B.C.-771B.C.), China's clothing system has gradually improved. There were imperial officials called Master of the Wardrobe (*Sifu*).All garments for emperors and ministers were custom-tailored, such as ceremonial robe, court dress, official dress and mourning clothes.

　　In accordance with the provisions of Western Zhou, when emperor, grand councilors or scholar-bureaucrats offered sacrifices to gods or ancestors, they shall wear ceremonial robes and crowns. Ranks were distinguished by the pattern and the quantity of tassels on crown. For example, there were twelve-ornament

冠圈：位于冕冠下方，用铁丝、竹藤、漆纱等编织成筒状。冠圈两侧开有小孔，称为"纽"。

Ring on the crown: vine and gauze which are knitted into cylinder shape by wires is beneath the crown. There are holes on each side of the rings called *Niu*.

冕板：冕冠顶部的盖板，上涂黑，下涂红，象征天与地。冕板略向前倾斜，象征天子勤政爱民。

Plate on the crown: Upper side of plate is painted red and lower side is in black, which represents the heaven and earth respectively. Plate is placed on the top of the crown which tilts forward a little bit to represent the philosophy of diligent emperor caring for his people.

玉笄：插在纽中，以便将冠固定在发髻上。

Jade hairpin: it's inserted into *Niu* to fasten crown on hair buns.

冕旒：冕冠的前后檐，垂有若干串珠玉，用彩线穿起来，称为"冕旒"。

Tassels: beads dangling on the front and rear brim of the crown are called tassels.

紞：冠沿、两耳附近各结一段彩色丝绳，叫做"紞"。有人认为其发展为天河带，即从冕板上垂下过膝的丝带，在隋唐以后的画中可以见到。

Silk ribbon: There are colored ribbons on the rim of the crown closed to ears called *Dan*. One deems that it evolved into Milky ribbons, long dangling ribbons to knees, which can be found in the portraits after Sui and Tang Dynasties.

舄：古代的鞋履与衣服一样，有着严格的礼制规定，在各种鞋履中，以舄最为贵重。舄是古代君王后妃以及公卿百官行礼时所穿的鞋，分赤、白、黑三色，以赤色为上。天子在最隆重的祭祀场合，身穿冕服，脚穿赤舄。

Clogs: There were also strict regulations on shoes in ancient times. Clogs are the most valuable among all kinds of shoes. When the emperor, empress and officials gave a salute, they wore red, white or black clogs. Red was the most supreme among three colors. Emperor wore ceremonial robe and red clogs to offer sacrifices to gods and ancestors, which was seen as the most important occasions.

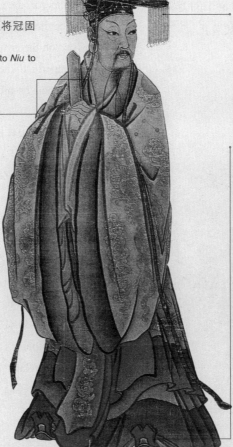

● 夏朝禹王冕服像
Portrait of King Dayu of Xia Dynasty in ceremonial robe

递减，如帝王用十二章纹，公卿用九章、九旒，侯伯用七章、七旒。

冕服作为传统的礼服世代沿用，不断有所变革，但基本形制未变。从西周至唐宋，天子、公卿、士大夫祭祀时皆穿冕服，但到了明代，冕服则成为帝王的专属，直到清代冕服制度才被废止。

and twelve tassels on the emperor's ceremonial robe. The emperor shall wear it on the grandest occasions. Its quantity shall be reduced accordingly on other less important occasions. When grand councilors and dukes attended ceremony to offer sacrifices to gods with the emperor, the embroidered patterns on their ceremonial robes and the number of tassels on the crowns shall be fewer than the emperor's. The emperor's had twelve; theirs shall be reduced to nine or seven accordingly.

Ceremonial robe as the traditional formal dress has been passed down from Western Zhou Dynasty to Tang and Song Dynasties (618-1279) with constantly innovation but little change in the basic form. During that time the emperor, ministries and scholar-bureaucrats wore ceremonial robe to offer sacrifices to ancestors and gods. In Ming Dynasty (1368-1644), it had been for the emperor's exclusive use until the institution related to ceremonial robe was abolished in Qing Dynasty.

• 晋武帝冕服像
Portrait of Emperor Wu of Jin Dynasty wearing ceremonial robe

十二章纹
Twelve-ornament Pattern

古代天子的冕服上绘绣十二种纹样，称"十二章纹"。衣上绘日、月、星辰、山、龙、华虫，称"上六章"；裳上绣宗彝、藻、火、粉米、黼、黻，称"下六章"。

Twelve-ornament is embroidered on the emperor's ceremonial robe. Six-ornament is embroidered on upper garment, including the patterns of sun, moon, star, rock, dragon and pheasant. The other six-ornament is embroidered on the lower garment. They are the pattern of drinking vessel for offering sacrifices to gods or ancestors, the pattern of algae, the pattern of flame, rice-shaped pattern, square patch embroidered with white and black axes and embroidery in square pattern.

日：太阳图案，绣于上衣的左肩处，与右肩的月亮遥遥相对，取光明照耀之意。

Sun: the pattern of the sun is embroidered on left shoulder of the upper garment, just opposite to the pattern of the moon on right shoulder, meaning bright shining.

月：月亮图案，绣于礼服的右肩，与日相对，取光明照耀之意。

Moon: the pattern of the moon is embroidered on right shoulder of the formal costume, opposite to the pattern of the sun, meaning bright shining.

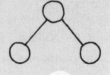

星辰：星象图案，通常用北斗七星，绣于日、月图案之下，或绣于后背，取光明照临之意。

Star: the pattern of the Galaxy, Charles's Wain is usually embroidered under the design of the sun and the moon or on the back, meaning brightness coming.

山：山石图案，绣在上衣，隐喻稳重、镇定。

Rock: the pattern of the rock is embroidered on the upper garment, meaning calm.

龙：双龙纹样，一只龙向上，另一只龙向下，取龙应变之意。

Dragon: the pattern of two dragons. One dragon goes up and the other goes down, meaning capability of handling emergency.

华虫：雉图案，因色彩鲜艳，纹章华丽而得名，表示穿用者有文章之德。

Huachong (Pattern of pheasant): beautiful ornament with bright color, which is embroidered on the robe, meaning the wearer with good virtues.

宗彝：祭器图纹，通常左右各一个，形成一对，器物表面常以虎、蜼为图饰（相传虎威猛而蜼智孝），取其忠孝之意。

Zongyi (drinking vessel for offering sacrifices to gods and ancestors): They are usually shown as a pair, one is on the left and the other is on the right. The designs on the drinking vessels are the tiger and golden hair monkey, meaning filial piety and loyalty.

藻：水草纹，常位于下裳，取其纯净、有文采之意。

Algae: It's embroidered on the lower garment meaning purity.

火：火焰图案，取其光明之意。

Flame: the pattern of flame meaning brightness.

粉米：米点状的白色花纹，取其滋养化育之意。

Fenmi: rice-shaped pattern, meaning nourishing.

黼：斧型纹样，斧刃为半白，斧身为半黑，取其决断是非之意。

Fu: Square patch on official costume is embroidered with white and black axes, meaning judging the right and wrong.

黻：两"弓"相背的纹样，用色为半青、半黑相间，取其明辨之意。

Foo: an embroidery in square pattern on official gowns with black and blue figure, meaning clear discrimination.

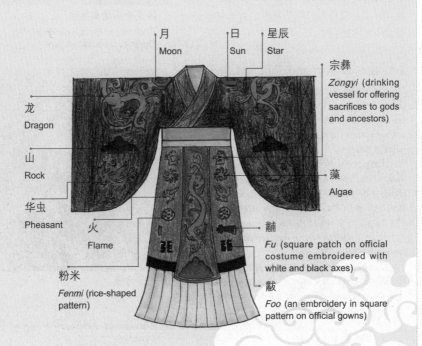

- 汉代皇帝的冕服
  Emperor's ceremonial robe of Han Dynasty (206B.C.-220A.D.)

冕冠和通天冠

冠是一种首服，在中国古代具有特殊的意义，是"昭名分，辨等威"的工具。古代贵族男子成年后就要戴冠，普通人则裹头巾。戴冠者，还要根据等级和身份的不同，佩戴不同形制的冠。

冕冠始于西周，是中国古代最重要的冠式，是帝王、王公、卿和大夫参加祭祀典礼时所戴的等级最高的礼冠。冕冠制度一直为后代沿用，冕冠的基本式样也被历代沿用。明代以后，冕冠被废，代之以朝冠。

通天冠也是中国古代皇帝的礼冠，但从秦代开始一般作为皇帝的常服，因此是级位仅次于冕冠的冠帽。

Emperor Crown and *Tongtian* Crown

Crown had special significance in ancient China. It's been a tool for "manifesting the social status". Nobel adult male shall wear crown, the ordinary with turban. Crown varied in different ranks of people.

Emperor's crown as the most important one in ancient China began in Western Zhou Dynasty (1046B.C.-771B.C.). The emperor, dukes, ministers and officials wore it to offer sacrifices. The basic style was followed in the following dynasties. Emperor crown was replaced by court crown after Ming Dynasty.

Tongtian (exceedingly high) crown was one of the emperor's crowns. Starting from Qin Dynasty it became emperor's informal crown, next only to ceremonial crown.

- **隋文帝头戴冕冠像**
Portrait of Emperor Wen of Sui Dynasty wearing crown

- **晋元帝像**
晋元帝戴着通天冠，上插玉簪，十分威严。
Portrait of Emperor Yuan of Jin Dynasty
Emperor Yuan of Jin Dynasty wearing *Tongtian* crown with jade hairpin looks very majestic.

> 龙袍

龙袍，又称"黄袍"，是中国古代帝王的袍服。黄袍作为帝王的专用服装源于唐代。唐代以前，黄色服饰在中国一直都较为流行，普通民众也可穿着。唐代以后，黄色

> Dragon Robe

Dragon robe also named as "yellow robe" was the gown-style dress of the ancient emperors. Starting from Tang Dynasty (618-907), it's for emperor's exclusive use. Yellow-colored clothes were popular and common people were allowed to wear before Tang Dynasty (618-907) but it became the royal costume after Tang Dynasty.

Yellow robe is also known as "Dragon robe" because there's embroidered dragon design on it. Dragon is a fictitious creature created by ancient Chinese. According to the legend, dragon has a snake-like body, fish scales, antlers and hawk talons. As the head of all beasts, it has the capability of overturning rivers and seas and calling up

• 唐太宗着黄袍像
Portrait of Emperor Taizong of Tang in dragon robe

● 明太祖朱元璋像

这张画像上，明太祖所穿的是衮龙服，为圆领的黄色袍服，绣有龙纹，穿时配翼善冠。

Portrait of Emperor Zhu Yuanzhang, the first founder of Ming Dynasty

Emperor Zhu Yuanzhang wearing round-collared yellow dragon robe with embroidered dragon design matching *Yishan* crown.

变为皇族的专用色彩，黄袍也成为皇族的专有服饰。

　　黄袍之所以又被称为"龙袍"，是因为上面绣有龙形图案。龙是古代中国人创造的一种虚拟动物，乃万兽之首。传说龙身长若蛇，有鳞似鱼，有角仿鹿，有爪似鹰，能走能飞，能大能小，能隐能现，甚至能翻江倒海，吞风吐雾，兴云降雨。在中国古代社会，龙是帝王的象征，凡是皇帝君王的器物，大都以龙为纹样。

　　中国历史上第一个身穿黄袍的皇帝为唐高祖李渊。他不仅以黄袍

the wind and invoking the rain. Dragon emblematizes the emperor in ancient China. Any utensils used by the emperor have the pattern of dragon.

　　Emperor Liyuan of Tang Dynasty was the first emperor who wore Yellow robe. He wore it in daily life and banned common people to wear yellow clothes. It's reiterated by Emperor Gaozong of Tang "All were forbidden to wear yellow dress except for emperors".

　　Traditional Chinese culture accounted for the reason why Emperor Liyuan of Tang decided on yellow-colored robe. In ancient China when the first emperor of a new dynasty was

为常服，还下令禁止庶民（普通百姓）穿黄色的衣服。到唐高宗时，更是重申"一切不许着黄"。

李渊之所以选择穿黄色的袍服，与中国传统文化有很大关系。在中国古代，开国皇帝登基时，为了给自己建立的王朝寻找统治根据，总要遵循"终始五德说"。"终始五德说"是战国时期阴阳家邹衍主张的一种观点。邹衍认为，历史上的每一个朝代都以土、木、金、火、水这五德的顺序交替统治，周而复始。而唐朝按"终始五德说"属"土"，土的代表颜色为黄色，因此李渊选择着黄袍。

从唐朝皇帝首穿黄袍开始，黄色正式成为皇权的象征。宋太祖赵匡胤"黄袍加身"后，更是禁止任何人穿黄袍，否则便以谋反论处。

crowned, he had to trace the ruling foundation for his dynasty, that is, following the opinion "The doctrine of five-virtue circulation" raised by an *Yin-yang* expert Zouyan who believed that each dynasty was ruled by five-virtue in order - earth, wood, gold, fire and water following one after another in rotation. Following this doctrine, Tang Dynasty was established in "earth" of five-virtue circulation which was represented by yellow, thus emperor Liyuan of Tang decided on Yellow robe.

Ever since Yellow robe was firstly worn by the emperor in Tang Dynasty (618-907), it became a symbol of imperial power. After "Emperor Taizu of Song in Yellow robe", anyone who wore Yellow robe would be punished for treason. Emperor Renzong of Song stated: common people were not allowed

- 翼善冠：古代帝王常戴的一种首服，以鎏金纱制成，上面附有双龙戏珠装饰。
  Yishan Crown: yellow dragon robe, the design of two dragons playing with a pearl worn by the ancient emperors.

宋仁宗时又规定：百姓着装不许以黄袍为底或以黄色配制纹样。从此，黄袍为皇室所独有，黄色亦为皇室所专用。这种制度为此后中国历代皇朝所沿袭，直至清朝灭亡。

to wear yellow gown or any dress with yellow pattern." Since then, yellow color and Yellow robe were the exclusive use for the royal family. This system has been followed until the demise of Qing Dynasty (1644-1911).

宋太祖"黄袍加身"

宋太祖赵匡胤在后周时期（951-960）只是一个手握兵权的将军。公元960年，辽国攻打后周边境，赵匡胤率兵抵抗，跟随他的还有弟弟赵匡义和亲信谋士赵普。当大军来到距京城二十里的陈桥驿时，赵匡胤命令将士就地扎营休息。当晚，一些将领聚集在一起，暗中商量拥护赵匡胤做皇帝。第二天一早，赵匡胤刚刚起床，下人就将早已准备好的一件黄袍穿在赵匡胤身上，然后众人跪倒在地，高呼"万岁"。就这样，赵匡胤率兵回到京城，顺理成章地当了皇帝，建立了宋朝。一件象征皇权的黄袍，一次精心策划的政变，使赵匡胤摇身一变成为大宋王朝的开国皇帝，足可见这件黄袍的意义和地位。

The First Emperor of Song in Yellow Robe

The First Emperor of Song, Zhao Kuangyin, was no more than a general having the military leadership during post Zhou Dynasty (951-960). When the Liao Kingdom launched a war to the border of post Zhou, followed by his brother, his trusted follower and counselor Zhao Pu, Zhao Kuangyin led his troop to defend against Jin. Zhao Kuangyin had his troop pitch at Chenqiaoyi which was about 10km away from the capital of the country. At that night some military officers gathered together to discuss secretly about advocating Zhao Kuangyin to be their emperor. When Zhao Kuangyin just rose the next day morning, he was dressed a ready dragon robe by his followers. Then all his followers kneeled down and shouted in chorus "Long live emperor". It's only natural that he became the emperor establishing Song Dynasty when he returned to the capital of the country. Zhao Kuangyin suddenly changed to the first emperor of Song Dynasty by a well-planned coup and a Yellow robe. Yellow robe which represented the imperial power had its great significance.

- 宋太祖常服像
  Portrait of the First Emperor of Song in informal dress

清光绪皇帝夏季朝服像
Portrait of Emperor Guangxu of Qing Dynasty in summer dragon robe

马蹄袖：马蹄袖本是北方少数民族射手服装的袖式。袖身窄小，紧裹手臂，袖口裁为弧线，袖口上半部可以覆盖在手背上。因为便于射箭，也称"箭袖"。清军入关后，将箭袖用于礼服之上，因外形看起来像马蹄，故名。

Horse-hoof sleeve: Its style originally comes from the northern minorities. Sleeve tends to be narrow wrapped the hand tightly. Cuff is tailored in arc shape and upper cuff covers the back of the hand for the convenience of shooting, also called "arrow sleeve". After Qing troop enter Shanhaiguan, formal attire adopts the design of "arrow sleeve". The shape of sleeve looks like a horse hoof, hence its name.

夏朝冠
Summer crown

夏披领：披领是清代帝后朝服使用的一种领饰，一般以绸缎为料，裁剪成菱形，上绣龙蟒等图纹，并加以缘饰。披领常缝缀在衣上，也有的与衣身分开，使用时罩在肩上，在颈部扣结。

Summer cape: It's a kind of decoration on the collar of court dress in the later period of Qing Dynasty. It's usually tailored in diamond shape with embroidered dragons and rims as decoration. It's made of silk and satin with dragon and python design. Cape is sometimes sewed on the dress. Sometimes it's separated from the dress and buckled on the neck end.

夏朝服龙袍裙
Summer court dress

清乾隆皇帝冬季朝服像
Portrait of Emperor Qianlong of Qing Dynasty in winter court dress

> 凤冠

中国古代戴冠者不限于男子，女子也戴冠，但只有王后、嫔妃、命妇（古代官员的母、妻等家眷）等身份地位较高者戴，普通女子的头部装饰则以发髻为主。汉代以前，贵族女子在参加祭祀典礼时，要戴假髻，上面安插金银珠翠饰品。从汉代开始，以凤凰饰首的风气在上层社会流行起来，凤凰形的冠随之产生，并逐渐发展成后来的凤冠。

> Phoenix Coronet

In ancient China, wearing crowns was not limited to men, women did too. But only for those who had the upper social status like the queen, imperial concubines, female families of government officials had the right to wear, while common women had hair buns for decoration. Before Han Dynasty (206B.C.-220A.D.), aristocratic women wore fake buns inserted golden, silver and emerald jewelry when they participated in sacrifice-offering ceremonies. Starting from Han Dynasty, phoenix-shaped ornaments became popular. Thus, phoenix-shaped crown came into being leading to the evolution of later phoenix coronet.

凤凰

凤和龙一样，也是一种想象出来的神瑞象征，是多种祥瑞动物的集合体。凤凰的形象一般为鸡首、燕颔、蛇颈、鹰爪、鱼尾、龟背和孔雀毛，羽毛一般为赤红色。在中国传统文化中，凤凰是首屈一指的吉祥瑞鸟。传说凤凰每次死后，周身都会燃起大火，然后在烈火中重生，并获得较以前更强大的生命力，称之为"凤凰涅槃"。如此周而复始，凤凰便会获得永生。秦始皇曾令三妃九嫔头插凤钗，足登凤头鞋，首次以政治力量把凤和女性用的服饰联系起来。

Phoenix

Just like dragon, phoenix is also an imaginary creature, a combination of many kinds of auspicious creatures. Phoenix image is composed of roster head, swallow chin, neck of a snake, talon of an eagle, tail of a fish, back of a turtle and crimson feather of a cocktail. Phoenix is the most auspicious bird in traditional Chinese culture. As a legend goes, phoenix comes back to life in the raging flames after its passing away and it will gain stronger life in each renascence. It's called "phoenix nirvana". It will gain the eternal life from this cycle. The first Emperor of Qin has ever ordered his concubines to wear phoenix-shaped hairpins and shoes, which was the first time in Chinese history to bring political power into relations between phoenix and women costume.

明代金凤凰饰品
Golden phoenix ornament of Ming Dynasty (1368-1644)

金龙装饰：冠顶部饰有三条金丝编制的金龙，其中左右两条口衔珠宝流苏。

Golden dragons as ornaments: three golden dragons woven by spun gold are on the top of coronet. The left one and the right one have beads in the mouth with tassels.

博鬓：凤冠后部饰有六扇珍珠、宝石制成的"博鬓"，呈扇形左右分开。

Bobin: Six panels of pearls and jewels at the rear of the phoenix coronet fan out in the right and the left directions.

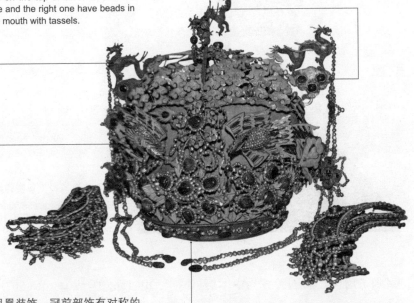

凤凰装饰：冠前部饰有对称的翠蓝色飞凤一对，凤背部满布珍珠，口衔珠滴。

Phoenix as ornaments: There is a symmetrical pair of flying phoenixes colored emerald blue in the front of the coronet. The back of the phoenix is covered with pearls. There's a bead in its beak.

宝石装饰：冠口沿镶嵌红宝石组成的花朵。

Jewels ornaments: The rim of the coronet is inlaid with ruby-made flowers.

材质：凤冠以髹漆细竹丝编制作胎，通体饰翠鸟羽毛点翠的如意云片，下缘镶金口。18朵以珍珠、宝石所制的梅花环绕其间。

Material: Mould of coronet is made of bamboo filament and lacquer. It's covered by kingfisher feather and dotted with cloud-shaped jade. Mounted gold is on the rim. There is 18 pearl-made and jewel-made plum blossoms around the coronet.

- 明孝端皇后凤冠
  Phoenix coronet for Empress Xiaoduan of Ming Dynasty (1564-1620)

明孝靖皇后着凤冠圆领袍服像
Portrait of Empress Xiaojing of Ming Dynasty (1565-1611) in the phoenix coronet and gown

到了唐代，武则天成为中国历史上的第一位女皇帝，凤凰也因此成为唐朝第一神鸟。因此在唐代遗物中，戴凤冠的女子形象很多。从宋代开始，凤冠正式纳入衣冠服饰制度，成为贵族女子的礼冠，后妃在受册或朝贺等隆重典礼时都要按规定戴上凤冠。此外，宋代还规定除皇后、妃嫔、命妇，其他人未经允许不得私戴凤冠。到了元代，后妃、命妇参加典礼时一般不戴凤冠，而是戴一种叫作"顾姑冠"的

By Tang Dynasty (618-907) when Empress Wu Zetian became the first female ruler in China's history, phoenix became the No. 1 magic bird of Tang Dynasty. Therefore, lots of phoenix coronets were discovered in the relics of Tang Dynasty. Beginning from Song Dynasty (960-1279), phoenix coronet was brought into the institutions of crown and costume. It has become the formal coronet for aristocrat. Concubines shall wear phoenix coronet in grand events of the court based on regulations. In addition, other than empress, concubines,

礼冠，具有鲜明的时代特色。

明代沿袭凤冠制度，皇后在接受册封或参加祭祀、朝会时，一定要戴凤冠。此外，明代嫔妃跟随皇帝参加祭祀和朝会时也要戴凤冠，但凤冠的形制和皇后的相比有所差异。主要是取掉了冠上的金龙，而代之以九翚鸟以示区别。明代礼制规定，命妇的礼冠只能用花钗、珠翠等，不能用凤凰。不过仍有一些达官贵族为了炫耀自己的地位和财

- 明孝端皇后像
  Portrait of Empress Xiaoduan of Ming Dynasty

female families of the government officials, other people were prohibited to wear phoenix coronet without permission. In Yuan Dynasty (1271-1368), concubines and women who were given the ranks by the emperor wore *Gugu* coronet instead when attending ceremonies, which bore distinctive feature of the times.

Following the coronet system of Ming Dynasty (1368-1644), empress in the participation of canonization, sacrifice-offering ceremonies or meeting with the emperor at court shall wear phoenix coronet. Concubines shall wear phoenix coronets when participating in some ceremonies with the emperor but theirs were quite different from the empress'. Difference lied in the pattern of the coronets. The golden dragon on concubines' coronets were removed and replaced by nine-*Hui* birds. Some aristocrats bought all kinds of phoenix coronets for their mother and wives to show off their wealth. The court found it hard to interfere with this issue because of its popularity. As time passed by, formal coronet was referred to as "phoenix coronet". With bud of capitalism in southern China, public wealth was occupied by individual. When phoenix

富，给自己的母亲或妻子置办各种各样的凤冠。由于这种情况十分普遍，朝廷很难干涉，时间久了，人们也将命妇的礼冠统称为"凤冠"了。随着南方资本主义的萌芽，社会财富大量被个人占有，凤冠在明代时逐渐流入民间，成为富家女子婚礼配套的服饰，新娘戴凤冠穿霞帔也成为人们认同其明媒正娶的重要符号。

coronet flew into the folk gradually, it became the wedding ornaments for wealthy family. Wearing phoenix coronet and *Xiapei* (a kind of shawl with opposite front pieces, big collar and slit below each flank of the crotch) was seen as an important symbol of a right and legal marriage.

After the founding of Qing Dynasty (1644-1911), some of China's traditional

- **元世祖皇后像**

  这幅元世祖皇后彻伯尔的盛装半身像，头戴珍珠饰顾姑冠，穿交领织金锦袍，雍容华贵。顾姑冠是宋元时期蒙古族贵妇所戴的一种礼冠。一般以铁丝、桦木或柳枝为骨，形体较长，女子戴着它出入营帐或乘坐车辇时，必须将顶饰取下。冠外裱皮纸绒绢，冠顶插有若干细枝，所用饰物均依戴冠者的身份等级而定，有金箔珠花、绒球、彩帛、珠串或翎枝，行动时摇曳生姿。

  Kit-Kat of Empress of the First Emperor of Yuan Dynasty

  In the kit-kat Empress Cheboer wears *Gugu* coronet with pearls, V-collar spun gold garment. She looks graceful. *Gugu* coronet is the ceremonial coronet worn by female Mongolia aristocrats in Song and Yuan Dynasties. It shall be taken off when wearers go into the tents and carriers. *Gugu* coronet is in long shape and the framework is made of wire, birch and swallow with silk and withes mounted on the top. Wearers' social status is distinguished by such ornaments as gold foil beads, pompons, colored silks, feathers and strings of beads. Shimmering coronets make wearers more graceful.

"八宝平水"纹：中国传统的织绣纹样，"八宝"包括犀角、宝珠、珊瑚、灵芝等。

Eight Treasures Embroidery Pattern: one of China's traditional embroidery patterns. "Eight treasures" includes the design of rhinoceros horn, beads, coral and glossy ganoderma lucidum etc.

领约：戴于颈间的一种装饰品。用金丝做成颈圈，上面镶嵌各种宝石，两端各垂有一条丝绦，金圈中装有可以开合的铰具，戴时打开套入颈间，压于朝珠和披领之上。

Neck strap: a kind of ornament around the neck. It's made of spun gold inlaid with different jewels, a fringe on each side. There is a hinge on the neck strap. It's pressed on the top of cape and court beads.

朝珠
Court beads

冬朝冠
Court coronet for winter

貂缘披领
Mink rim collar

金约：朝冠的配件。在戴朝冠时需先戴金约，再戴朝冠，起到束发的作用。

Jinyue: supporting part of the court coronet. Wearing *Jinyue* first to make hairdo, then wearing the coronet.

行龙纹：中国传统的织绣纹样，龙身侧向，昂首竖尾，龙爪向下，表示行走。有别于正面端坐的"正龙"，经常对称使用，左右各一。常刺绣在衣服的腰及下摆部位。

Walking Dragon Pattern: It's one of China's traditional embroidery patterns. Walking dragon presents walking image of its profile, raising its head, erecting its tail, its claw downwards. Its image is different from the seated "front-side-view dragon" which is embroidered below the waist part of the gown one on each side.

貂缘马蹄袖
Hoof-shaped cuff with mink rim

彩帨：系挂在衣襟处，以彩帛为料，做成狭长的条状，上窄下宽，底部为锥形，用颜色与织绣的纹样来区分上下等级。

Caihui (kerchief): It's attached to the front piece of the garment. It's in a conical shape with wide bottom and narrow top. People's social status is distinguished by its color and pattern.

• 清孝贤皇后冬季朝服像
Portrait of Empress Xiaoxian of Qing Dynasty (1712-1748) in winter court dress

清王朝建立后，中国的传统服制尽数被废，但以凤凰饰首的习俗得到了保留。清代后妃参加庆典，都戴一种折檐软帽，称为"朝冠"，事实上也是一种凤冠。

costume institutions were abolished; however, the phoenix coronet retained. Concubines in Qing Dynasty wore a kind of floppy crown with folded rim, called "court coronet". It's one of the phoenix coronets.

- **清代金凤冠**

清代的凤冠上覆有红色丝纬，在丝纬的四周，缀有7只金凤。在帽子正中，还叠压着3只金凤，每只金凤的顶部，各饰一颗珍珠。

Golden phoenix coronet of Qing Dynasty (1644-1911)

It's coated with red wires. There are seven gold phoenixes around wires. Three golden phoenixes are overlapped in the center of the coronet. There is a pearl on the top of each golden phoenix.

- **清代皇后冬季朝冠**

Empress of Qing Dynasty in winter court coronet

朝服与官服
Court Costume and Official Costume

从西周开始，中国的等级制度逐步确立。在服饰方面，周王朝设"司服""内司服"等官职，掌管王室服饰。王室公卿为表尊贵威严，在不同的礼仪场合，穿衣着裳也须采用不同的形式、颜色和图案。朝中官员的服饰也有严格规定，要根据官职的大小制定服饰的品级。

China's hierarchy was gradually set up from Western Zhou Dynasty (1046B.C.–771B.C.). The position of Master of Wardrobe (*Sifu*) and Eunuch Master of Wardrobe (*Neisifu*) were in charge of imperial costumes. Imperial members wore different costumes on different occasions with various designs and colors to show their majesty. At that time, they also had strict regulations on officials' costumes. Costumes were broken down by officials' ranks in the court.

> 朝服

朝服是中国古代帝王、百官及后妃、命妇参加朝会时所穿的礼服，由祭服演变而来。根据文献记载，早在西周时就已有朝服，称为"皮弁"，一直沿用至明代。皮弁服的首服采用白鹿皮制成，上衣为细布白衣，下为素裳。因衣裳朴素无纹，所以在皮弁上装饰玉石，以玉色、数量区分身份等级。官员回到家中，如果要会见比自己身份低的人，则不能再着皮弁服，而要换上一种用黑布制成的称为"缁衣"的朝服。

春秋战国时期的朝服多用黑色布帛制成，称为"玄端"。西汉时期的朝会之服也用黑色，称为"皂衣"。东汉时期，上至帝王，下至小吏，皆以袍作为朝服。此时的袍

> Court Costume

Court dress evolving from sacrificial dress was formal attires for emperors, all officials, empresses, imperial concubines in ancient China. According to literature, court dress appeared as early as Western Zhou Dynasty (1046B.C.-771B.C.), known as *Pibian* which has been in use till the Ming Dynasty (1368-1644). The crown for *Pibian* was made of white deer hide. Dress for upper part of the body was made of white fine cloth, lower dress in plain. Considering there was no pattern on lower dress, jades were added on *Pibian*. People's social status was judged by color and the quantity of the jade. *Pibian* was not allowed to wear when one met a person who ranked lower than him. He should change a black court dress, known as *Ziyi*.

In the Spring and Autumn and Warring States Period (770B.C.-221B.

唐代官员朝服像
Portrait of officials in court dress of Tang Dynasty (618-907)

服为上衣下裳不分的深衣制，因所用质料多为绛色纱，故称"绛纱袍"。帝王、百官穿袍服时，还要腰系大带、革带、佩绶、佩玉等。东汉的朝服制度一直沿用至宋、明。事实上，朝服在演变过程中，已渐渐成为官服的一种。

清代的百官朝服品级差别更为鲜明，主要在于冠服顶子和蟒袍的纹饰。朝冠后插有翎枝，其制六品以下用蓝翎，五品以上用花翎；百官的蟒袍，一品至三品绣五爪九

C.), court dress was mainly made of black cloth and silk, known as black clothes (*Xuanduan*). During Western Han period (206 B.C.-25A.D.), officials wore it when having meetings in court, called dark coat (*Zaoyi*). Gown-style dress became court dress in Eastern Han Dynasty (25-220). This dress was an unlined garment. Because of its dark red color and material, it's called Red Satin Gown (*Jiangsha Pao*). Emperors and officials wore gowns with leather-made band, ribbons and jades. Court dress system of Eastern Han Dynasty has been used

- 绛纱袍

  绛纱袍为深红色纱袍，交领大袖，下长及膝，领、袖、襟等皆滚以黑边。

  ### Red satin gown (*Jiangsha* gown)

  It's a dark-red satin gown down to knees with crossed collar and big sleeves. Rims of collar, front piece and sleeves are in black.

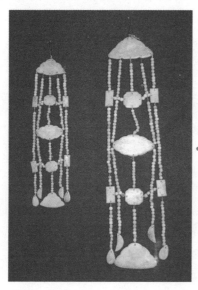

- 明代朝服的玉佩

  此玉佩称为"大佩"，是玉佩中最为重要的一种。使用时将其悬挂在腰下，专门用于贵族男女的祭服或朝服，始于商周，历代沿袭，入清之后废止。

  ### Jade pendants for court dress of Ming Dynasty (1368-1644)

  This jade pendant known as "grand pendant" is the most essential one. Starting from Shang and Zhou Dynasties (1600B.C.-256B.C.), it's been used to match the sacrificial dress and court dress of the aristocratic men and women until it's abolished at the beginning of Qing Dynasty.

蟒，四品至六品绣四爪八蟒，七品至九品绣四爪五蟒；文官五品、武官四品以上及科道、侍卫等职，均需悬挂朝珠。

- 明代权臣严嵩朝服像
Portrait of powerful minister Yan Song of Ming Dynasty in court dress

until Song and Ming Dynasties (960-1644). In the process of evolvement, it has gradually become one of the official costumes.

In Qing Dynasty (1644-1911), significant difference in ranks of court dress could be found in ornament patterns on gowns and caps. Feathers were inserted in the rear of the caps. Feathers of Brown Eared-pheasant dyed blue were used for officials rank below the sixth. Peacock feathers were used for others above the fifth. Officials of the first rank to the third rank wore robes embroidered with nine pythons with five claws. Eight pythons with four claws pattern was for the fourth to the sixth rank officials. For the seventh to the ninth ranks, the pattern was five pythons with four claws. Civil officials of the fifth rank and above, military officials of the fourth rank and above, Supervising Censors (*Kedao*) and Imperial Guards must wear court beads.

> 官服

官服，又称"公服"，是中国古代官吏或侍从在处理公务时所穿的服装。早期的官服名叫"袧衣"，是一种单衣，两袖窄小，便于从事公务，这也是有别于祭服、朝服之处。据史书记载，袧衣作为官吏所穿官服的主要式样，一直沿用至隋代。

到了唐代，官服制度较为完善，官服的形制采用袍制，两袖仍比较窄小，以服色、纹样、佩饰区分官吏的等级身份，对后世官服产生了深远的影响。宋代官服款式为圆领、大袖袍服，腰束革带。元代官服沿袭宋代，但又有所创新，在官服上绣以花卉图案，以图案品种、大小区分品级。明代的官服为袍式，盘领，右衽（衽即衣襟），

> **Official Costume**

Official costume, also known as *Gongfu*, was worn by ancient Chinese officials when they dealt with public affairs. *Gongfu* called *Gou Yi* in early times, was a garment of single layer with narrow sleeves, which was easy to engage in public affairs, different from the memorial ceremony dress and court dress. According to historical records, it had been the main clothing worn by officials until Sui Dynasty (589-618).

Official dress system in Tang Dynasty (618-907) tended to be perfect. The official dress was gown-style with narrow sleeves. Ranks of officials were distinguished by the color, pattern and accessories, which had profound influence for later improvement of the official dress. Official dress of Song Dynasty (960-1279) was gown-style with round collar, loose sleeves and girded

乌纱帽：明代的官帽，由唐宋时期的幞头演变而来，以铁丝为框，外蒙乌纱，帽身前高后低，左右各插一翅，文武百官上朝均可戴，入清以后其制被废。

Black gauze cap: the official cap of Ming Dynasty (1368-1644). It evolves from a kind hood called *Futou* of Tang and Song Dynasties. Its framework is made of wires and covered by black gauze. Rear part is lower than front part inserted a wing in the right and left side respectively. Civil and military officials are allowed to wear but it's abandoned by the beginning of Qing Dynasty (1644-1911).

官靴：明清时官靴的靴筒长过脚踝，至小腿的三分之一左右，靴首略翘，通常用黑色皮革或黑色绸缎为料制作，与官服配用。

Official boots: the length of the boots of Qing Dynasty reaches about one third of crura above the ankles. The toe of the boots tends to tilt upwards. Boots are usually made of black leather or silk and satin to match the official dress.

补服：补服是明清时期官员的官服，用于官员朝视、谢恩、礼见、宴会等场合穿用。因胸前、背后缀有补子，故名。

Bufu: It's the official dress of Ming and Qing Dynasties. It's worn on the occasion of court meeting, ritual of gratitude, ritual of meeting the emperor and banquet etc. Its name comes from the character *Bu*, a kind of ornament badge on both back and chest of the dress.

• 明代官吏像
Portrait of official of Ming Dynasty (1368-1644)

袖宽三尺，多用苎丝、纱、罗等材料制成。根据公服的服色、绣花的花种和大小以及腰带的材质区分品级。一至四品为绯色；五至七品为

waist with a waistband. Following the style of Song Dynasty, official dress had its innovation of adding the flower embroidery. The ranks of officials were seen from the category of embroidered

- 忠靖服

忠靖服是明代职官退朝闲居所穿着的服装，配忠靖冠，含"进则尽忠，退思补过"之意，交领，右衽，大袖，上下相连，衣长过膝，常以深青色纱罗为料。不同的品级官员以不同的图案或素色来区别地位等级。

Loyal peace (*Zhongjing*) dress

Loyal Peace Dress is the home garment of Ming Dynasty (1368-1644),together with Loyal Peace Hat. Loyal Peace means keeping forging ahead and mending mistakes. The length of the garment is over knees. It's made of dark cyan with crossed collar, large sleeves and right front piece. Ranks vary in different patterns and plain colors.

- 忠靖冠

忠靖冠是明代的官员退朝闲居时所戴的帽子。制作时用铁丝围成框架，用乌纱、乌绒包裹表面。冠的形状略呈方形，中间微突，前面部分装饰有冠梁，并且压有金线。后部的形状像两个小山峰。冠前的梁数根据官职的品级而定。

Loyal peace (*Zhongjing*) hat

Officials of Ming Dynasty wore Loyal Peace Hats (*Zhongjingguan*) at home. The framework of this kind hat is made of wires and covered by black gauze and black velvet. It's in square shape with projecture in the middle. There is a beam pressed by gold thread on the front part of the hat. The shape of the rear looks like two hills. The number of beams on front part is judged by the rank of the official.

- 清代官吏像

长袍马褂是清代有身份的男子，尤其是官吏非常典型的服装。图中的长袍为大襟，窄袖，袍长至脚踝，两侧或中间开衩。马褂为立领、对襟、窄袖，下摆较宽。

Portrait of the official in Qing Dynasty (1644-1911)

Long gown and ridding jacket are worn by men with high social status. It's a typical costume for officials. This illustration shows the front piece, narrow sleeves, slits on both sides or in the middle and ankle-length gown. Ridding jacket has stand collar, opposite front pieces, narrow sleeves and wide hem.

青色；八至九品为绿色。清代的官服废除了服色制度，不论职位高低，颜色都是蓝色，只在庆典时方可用绛色。清代官服由袍、褂组成，袍均为圆领，右衽。

官服的颜色品级

官服的颜色是区分官吏等级的标准之一。以唐代为例，唐贞观四年（630年），官服的颜色被定为四等：一品至三品服紫，四品至五品服绯，六品至七品服绿，八品至九

● 唐代身穿各色官服的侍从
Attendants in different official dresses of Tang Dynasty (618-907)

flower and its size. Official dress of Ming Dynasty (1368-1644) was gown-style woven by silk and yarn. It had rotating collar, right front piece and three-*Chi*-wide sleeve (one *Chi* equals to 33.33cm). Official ranks were distinguished by colors of the dress, sizes of the embroidered flowers and materials of waistband. From the first to the fourth ranks of officials, they had dark red dress; from the fifth to the seventh, they were in cyan; from the eighth to the ninth, they wore green dress. Color regulations on official dress were abolished in Qing Dynasty (1644-1911). Regardless of the ranks, all in blue, with the exception of celebrations, they wore dark red dress. Official dress of Qing Dynasty consisted of unlined garment and round-collar gown with right front piece.

Colors and Ranks of Official Costume

Colors of official costume were taken as one of the standards to distinguish official ranks. Taking Tang Dynasty as an example, the fourth year of Zhenguan of Tang in throne (630), colors of the official uniform were decided for four levels. Purple is used for the first to the third ranks. Red is for the fourth and the fifth.

品服青。唐末又规定：三品以上仍旧用紫色，四品用深绯，五品用浅绯，六品用深绿，七品用浅绿，八品用深青，九品用浅青。

由于紫色的官服在唐代最为尊贵，所以"紫袍"一词也成为显官要职的代称。绯色的官服即指大红色的袍服，大袖，右衽，衣襟及袖

Green is for the sixth and the seventh. Cyan is for the eighth and the ninth. In late Tang Dynasty these regulations were updated. The color of official costume for the third rank and above was purple unchanged. For the fourth was dark red, for the fifth was light red, for the sixth was dark green, for the seventh was light green, for the eighth was dark cyan and for the ninth was light cyan.

Since purple was seen as a noble color, people who wore purple robe were regarded as an important position in Tang Dynasty. Red gown had large sleeves, right front piece with edge piping on the cuff and collar. Green robe was for the sixth and the seventh rank of officials who were in low position, whereas

- 《杏园雅集图》（明 谢环）

此图描绘了明代官员集会的场景，图中的官吏均戴乌纱帽，身穿各色袍服，腰系玉带，是明代仕宦服装的真实写照。

Painting of gathering in apricot garden (by Xie Huan, Ming Dynasty)

This painting depicts the scene of officials' rally. All the government officials wear black gauze hats, all kinds of gowns and jade-sash. It presents a vivid description of official costume in Ming Dynasty.

口常有镶边。绿袍是唐代六品及七品官的官服，品级相对较低。青袍是唐代官服中等级最低的，后来"青袍"一词多用来代称品级低的官吏。

唐代的公服制度对服色的规定虽然很严格，但在具体实行的时候，也可以变通。如果一些官吏的品级不够，但遇到奉命出使等特殊情况，经过特许可穿用比原品级高一级的服色，俗称"借紫"或"借绯"。

officials with cyan gown ranked the lowest among all the moderate positions. Later on people referred to "cyan gown" as low-rank officials.

Even though there were strict rules for the official costume in Tang Dynasty, it tended to be flexible when these rules were put into practice. There were some lower ranked officials presented special occasions as being sent on a diplomatic mission, they were authorized to wear a higher ranking costume. It's referred to as "borrowing purple" or "borrowing red".

"衣冠禽兽"的来历

唐代官吏所穿的袍服纹样一般以暗花为多。武则天当上皇帝后，颁布了一项新的服装制度，即在不同职别官员的袍服上，绣以不同的图案，称为"绣袍"。文官的袍上绣飞禽，武官的袍上绣走兽，"衣冠禽兽"这个成语就源于此。在文武官员的袍服上绣禽兽纹样以区分官员品级，这一制度被后世所沿袭，明清时期发展成"补子"。

Origin of beast in human attire

The design for official gowns of Tang Dynasty was mostly dim flower patterns. Wu Zetian released a new edit when she came in throne. Known as Embroidered Gown, varied embroidery patterns were for different ranks. Costume with embroidered fowls was for civil official, and the one with embroidered beasts was for military official. It's where the idiom "beast in human attire" originated. It was a commendatory epithet falling into ill reputation. The patterns of fowls and beasts had been embroidered on the officials' costumes as a rule passed down to the following dynasties. It's developed into *Buzi* in Ming and Qing Dynasties.

绣袍：四名侍从身穿绣袍，站在唐明皇身旁。
Embroidered robe: Four attendants wearing the embroidered gowns are standing beside the Emperor Ming of Tang Dynasty.

黄袍：唐明皇身穿黄袍，头戴幞头，坐在龙椅上。
Yellow robe: Emperor Ming of Tang Dynasty is in yellow robe and *Futou* hat sitting on the dragon chair.

紫袍：这个官吏身穿紫袍，束腰带，头戴幞头，双手相握而立。
Purple robe: This official is in purple robe and *Futou* hat, standing there overlapping his hands.

- 《张果老见明皇图》（元 任仁发）
  Painting of Zhang Guolao meeting the Emperor Ming of Tang Dynasty(by Ren Renfa, Yuan Dynasty)

补子

补子是明清官服上的纹样图案，源于绣袍，一般用金线或彩丝绣织成，分为禽、兽两类图案，文官用禽，武官用兽，用来区分官员的身份等级。补子的外形通常做成方形，在袍服的前胸和后背各缀一块。

Buzi

Buzi, originated from the embroidered gown, was the pattern on the official dress of Ming and Qing Dynasties. Woven by golden thread or colored yarn, it fell into two categories, that is, the patterns of fowls for the civil officials and the pattern of beasts for the military officials to distinguish their positions. *Buzi* was mostly in square shape attached one piece to fore breast and back of the gown respectively.

明代文官补子：
***Buzi* on civil official's costume in Ming Dynasty (1368-1644)**

一品：仙鹤
The first rank: crane

二品：锦鸡
The second rank: golden pheasant

三品：孔雀
The third rank: peacock

四品：云雁
The fourth rank: wild goose

五品：白鹇
The fifth rank: silver pheasant

六品：鹭鸶
The sixth rank: egret

七品：鸂鶒
The seventh rank: Xi Chi (a kind of water bird)

八品：黄鹂
The eighth rank: yellow oriole

九品：鹌鹑
The ninth rank: quail

杂职：练雀
Other officials: finch

御史：獬豸
Ministry: *Xiezhi* (a legendary beast in ancient China)

明代武官补子：

Buzi on military official's costume of Ming Dynasty

一、二品：狮子

The first and second rank: lion

三品：虎

The third rank: tiger

四品：豹

The fourth rank: leopard

五品：熊

The fifth rank: bear

六、七品：彪

The sixth and seventh rank: young tiger

八品：犀牛

The eighth rank: rhinoceros

九品：海马

The ninth rank: hippocampi

清代文官补子：
Buzi on the civil official's costume of Qing Dynasty

一品：仙鹤
The first rank: crane

二品：锦鸡
The second rank: golden pheasant

三品：孔雀
The third rank: peacock

四品：云雁
The fourth rank: wild goose

五品：白鹇
The fifth rank: silver pheasant

六品：鹭鸶
The sixth rank: egret

七品：鸂鶒
The seventh rank: *Xi Chi* (a kind of water bird)

八品：鹌鹑
The eighth rank: quail

九品：练雀
The ninth rank: finch

御史：獬豸
Ministry: *Xiezhi* (a legendary beast in ancient China)

清代武官补子：

Buzi for the military officials' costume of Qing Dynasty

一品：麒麟

The firsth rank: kylin

二品：狮

The second rank: lion

三品：豹

The third rank: leopard

四品：虎

The fourth rank: tiger

五品：熊

The fifth rank: bear

六品：彪

The sixth rank: young tiger

七品、八品：犀牛

The seventh and eighth rank: rhinoceros

九品：海马

The ninth rank: hippocampi

- **清代官服上的补子**

  清代的补子基本沿承明代，但也有所变化：明代的补子施于袍上，而清代补子用于褂上；明服为团领衫，前胸补子为完整的一块，而清服是对襟褂，前胸的补子被一分为二；明代的补子大约40厘米见方，清代的补子稍小，约30厘米见方；明代的补子多以红色等为底，金线绣花，而清代的补子则是以青、黑、深红等深色为底，五彩织绣；明代的补子只饰于前胸后背，清代宗室的圆补有的不仅饰胸，还饰于两肩之上。

Buzi on the official dress of Qing Dynasty (1644-1911)

Buzi of Qing Dynasty basically evolves from Ming Dynasty with some changes. Dress of Ming Dynasty has a round collar. *Buzi* of Ming Dynasty is a whole piece of 40cm square cloth attached to fore breast of gown. It has flower embroidery with gold thread against a red background. Whereas the costume of Qing Dynasty is opposite front pieces jacket with two halves of *Buzi* which is 30cm square against a dark red, black and cyan background. Colorful embroidered *Buzi* is not limited to decorate fore breast and back of the dress, it's also on two shoulders.

官服的配饰品级

官服的配饰也是其品级的体现。以唐代为例，官吏按照品级高低，穿用不同材质的腰带，如金、玉、犀、银、石、铜、铁，并以腰带上的饰物区分等级。

中国古代官服的配饰有很多，其中最具代表性的就是佩绶。佩，指佩戴于身的玉饰，如大佩、组佩等；绶，指用来悬挂印、玉佩的丝

Accessories and Ranks of Official Costume

Official ranks were reflected by the accessories on their official costumes. Taking Tang Dynasty (618-907) as an example, officials wore waistband in different materials such as gold, jade, rhinoceros horn, silver, stone, copper and iron. Their positions were also distinguished by the ornaments on the sashes.

带。以佩绶区分尊卑是我国古代服饰制度的显著特征。佩绶在秦代以前就已出现，并作为一种官服制度流传下来，但绶带的色彩规格时有变化。到了清代，这种佩绶制度不再使用，取而代之的是顶戴制度。

玉佩也是中国古代贵族和官员们礼服上必不可少的一种装饰，受森严的等级约束，玉佩的形制、佩带的方法及部位也都据佩玉者的身份有明确的规定。

Ancient China boasted lots of accessories. *Peishou* (ribbons hanging jade ornaments) was the most representative one among all accessories. *Pei* referred to jade ornaments such as large jade and groups of jades. *Shou* referred to the ribbons hanging jades and seals. To distinguish hierarchy by *Peishou* had been a distinctive feature for the institution of ancient Chinese ornaments. *Peishou* appeared prior to Qin Dynasty. As an institution of ornament, it's passed down with some change in the color and shape. The institution related to *Peishou* wasn't terminated until hat wearing system appeared in Qing Dynasty (1644-1911).

In addition, as an essential ornament on formal attire for ancient aristocrats and officials, jade pendant was restricted by many—layered hierarchy. Wearers' identity, shape of jade pendant and the way to wear were clearly defined.

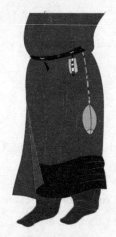

- **唐代官服上的鱼袋配饰**

  鱼袋是一种佩囊，用来盛放鱼符，系挂于腰部。鱼袋的颁发与鱼符相同，均由朝廷操办，宋代没有鱼符制度，但官员仍佩有鱼袋。

  Fish Bag (*Yudai*) ornament on official costume of Tang Dynasty (618-907)

  Fish Bag (*Yudai*), a kind of pocket for holding fish-shaped tallies is tied on the waist. The court awarded fish-shaped tallies and Fish Bag (*Yudai*). Institution related to fish-shaped tallies was not established in Song Dynasty but *Yudai* was available at that time.

- **唐代官服的皮革腰带**

  Leather belt on official costume of Tang Dynasty (618-907)

- 鱼符

鱼符是隋唐时朝廷颁发给官吏的鱼形符契，鱼符质料因官阶不同而有所区别。鱼符上面刻有文字，分成两爿，一爿在朝廷，一爿自带。官员迁升或是出入宫廷等情况，以鱼符吻合为凭信。

Fish-shaped tallies

The imperial court of Sui and Tang Dynasties issued fish-shaped tallies to officials. Material varies with ranks of officials. There are engraved characters on the fish-shape tallies. It falls into two slit bamboos or chopped wood. One half is kept by the court and the other half is kept by the official. Two halves have to be tallied with each other before entering and going out of the imperial court.

- 印绶

印绶是指系缚在印纽上的彩色丝带，一般打成回环。印绶的颜色和长度都有具体的规定，用来区分身份级别。汉代的官印被盛放在腰间的绶囊里，印绶佩挂在腰间，常垂搭在囊外或与官印一同放入囊内。

Yinshou

Yinshou is colored ribbon looped on the knob of a seal. There are regulations on *Yinshou*'s length and color to distinguish people's ranks. In Han Dynasty, the official's seal was put into seal bag (*Shounang*). *Yinshou* is hung on the waist either dangling outside or putting together with seal and silk ribbon into the seal bag.

- 双绶

双绶是穿着官服时佩戴的两条丝带，通常挂在腰部，左右各一条，由秦汉时期的印绶演变过来。南北朝以后广为流行。

Double *Shou* (double silk ribbons)

Double *Shou* refers to two silk ribbons to match official costume. Officials usually wear two silk ribbons hanging on each side of their waists. It evolved from *Yinshou* of Qin and Han Dynasties (221B.C.-220A.D.) and became popular after Northern and Southern Dynasties.

- 三绶

汉代的官员有一枚印章便佩戴一印绶，若有数枚印章，则配用数条印绶。绶越多，则表示官员的权力越大。

Triple *Shou* (three silk ribbons)

The official in Han Dynasty would wear a silk ribbon if he had one seal. The more seals he had, the more ribbons he wore. The more ribbons he was offered, the more authority he had.

宋代官服品级表
Rank of the official dress of Song Dynasty

品级 Rank	服色 Color	腰带 Sash	鱼袋 Fish Bag
一品 First-rank	紫色 Purple	玉带 Jade sash	金鱼袋 Goldfish bag
二品 Second-rank	紫色 Purple	玉带 Jade sash	金鱼袋 Goldfish bag
三品 Third-rank	紫色 Purple	玉带 Jade sash	金鱼袋 Goldfish bag
四品 Fourth-rank	紫色 Purple	金带 Jade sash	金鱼袋 Goldfish bag
五品 Fifth-rank	绯色 Red	金漆带 Gold lacquer sash	银鱼袋 Silverfish bag
六品 Sixth-rank	绯色 Red	金漆带 Gold lacquer sash	银鱼袋 Silverfish bag
七品 Seventh-rank	绿色 Green	黑银及犀角带 Black silver and rhinoceros horn sash	
八品 Eighth-rank	绿色 Green	黑银及犀角带 Black silver and rhinoceros horn sash	
九品 Ninth-rank	绿色 Green	黑银及犀角带 Black silver and rhinoceros horn sash	
庶人 Common people	皂白色 Black and white	铁角带 Iron and rhinoceros horn sash	

顶戴花翎

　　顶戴花翎是清代独特的礼帽，分两种，一为暖帽，一为凉帽。礼帽在顶珠下有翎管，质为白玉或翡翠，用来安插翎枝。翎枝又分为蓝翎和花翎两种，蓝翎为鹖羽（鹖：今名褐马鸡）所做，花翎为孔雀羽所做。蓝翎只赐予六品以下的官员或在皇宫和王府当差的侍卫佩戴，也可以赏赐建有军功的低级军官。花翎在清代是一种高品级的标志，非一般官员所能戴用，一般被罚拔去花翎的官员一定是犯了重罪。花翎又分一眼、双眼和三眼，以三眼最为尊贵。这里的"眼"是指孔雀翎上眼状的圆。花翎如此高贵，因此在清代非常受重视，享有的官员也极少，乾隆至清末期间被赐三眼花翎的大臣仅有七人，被赐双眼花翎的也只有二十多人。

Official Hat with tail feather

Official Hat with tail feather was a unique style in Qing Dynasty (1644-1911) included winter hat and summer hat. Feather duct (*Lingguan*) was made of jade and emerald jade (used as a holder for the feather) beneath the top bead. There were two types of feather, blue feather (feather of brown eared-pheasant dyed blue) and peacock feather. Blue feather was offered to the officials below the sixth rank. In addition, it's granted to guards serving in the imperial palace and the low rank military officials who performed meritorious deeds in battles. Peacock feather as a symbol of high rank was exclusive for high rank officials. Punishment of removing the peacock feather was regarded a felony.

- 暖帽

　　暖帽为官员冬季所戴，形状多为圆形，周围有一道朝上翻卷的檐边，檐边的材料根据气候的变化分为皮毛、呢绒等。在帽的顶部，一般还装有帽纬，帽纬之中又饰有顶珠。除翎枝外，暖帽顶珠的颜色及材料也是区分官职的重要标志。

Winter hat

Winter hat is worn in winter. It's a round hat with brim curling upwards. The material of brim varies in leather, wool and other fabrics for weather variations. The top of the hat is called *Maowei*. There is a bead on the *Maowei*. Besides the feather on the hat, the color and the material are chief signs to distinguish officials' ranks.

Peacock feathers were fallen into "one-eye", "two-eye" and "three-eye". Here "Eye" refers to the eye-shaped markings on the feather. Three-eyed was the most honorable. Peacock feather was so dignified that it was held in great honor in Qing Dynasty. Seldom officials have ever had this honor. Only seven ministers were bestowed three-eye peacock feather and twenty were granted two-eye peacock feather from Emperor Qianlong period to the end of Qing Dynasty.

- 凉帽

凉帽为官员夏季所戴，呈圆锥形，清初时崇尚扁而大，后流行高而小的形状。通常用藤竹、篾席、麦秸草编结帽体，外裱绫罗，内衬红色纱罗，沿口镶滚片金缘，顶部装饰红缨、顶珠、翎管和翎羽。根据清代礼冠制度，每年春季三月将暖帽换成凉帽，八月将凉帽换成暖帽。

Summer hat

It's a cone-shaped hat worn in summer time. It tends to be large and flat at the beginning of Qing Dynasty. Later the small and tall shape becomes popularity. its frame is woven by vines, thin bamboo strips and grass with mounted silk and red gauze lining. The hat has gold edge piping. The top of the hat is adorned with a bead, a feather and strips of red. According to the institution of Qing Dynasty, wearing summer hats started in March every year and winter hats were replaced by summer hats in August.

便服
Casual Wear

便服是指古代人平常家居时所穿的衣服。便服与礼服是相对的服装，礼服是礼制的产物，服装的质地、色彩、款式、纹样等多有统一而严格的规定，历代延续，变化不大。而便服是居家之服，主要受社会风尚和生活习俗等影响，受礼制约束比礼服少，所以形式多样、变化无穷。在中国历史上，一度流行的便服形制主要有深衣、襦、衫、袍、袄、褂等。

Ancient people wore casual wear at home. Compared with casual wear, formal attire was the outcome of social etiquette. There were strict regulations on its material, color, style and pattern. It's passed down from dynasty to dynasty with little change. Whereas the casual wear had little influence from etiquette. It's varied in style and pattern thanks to the influence of fashion and little restriction from a set of etiquette. Such casual wear as *Shenyi*, jacket, underlinen, gown, short coat and unlined garment gained ground in Chinese history.

> 深衣

　　深衣最早出现于春秋战国时期，最初只是士大夫阶层居家的便服，由于其形制简便，穿着适体，后来用途越来越广，最后无论男女，不分尊卑，皆可穿着。魏晋以后，深衣被袍衫所取代，才逐渐退出历史舞台。但后世的裤褶、襦裙等服装都以深衣为原型。

　　在深衣出现之前，古人的衣服都是由上衣和下裳组成的，而深衣却打破了这一风俗，上下连为一体，呈直筒式。不过，深衣的形制特征仍是上下分裁而合制，上下保持着一分为二的界限。

　　深衣的形制，每一部分都有其独特的寓意。比如在制作中，先将上衣下裳分裁，然后在腰部缝合，这是为了尊祖承古；采用圆袖方

> *Shenyi* (Garment)

Shenyi first appeared in the Spring and Autumn and Warring States Period (770B.C.-221B.C.). It was home-worn dress for scholar-bureaucrat initially. Owing to its simple design for comfortable wear, it was widely spread among common people. When *Shenyi* was replaced by gown-style dress, it gradually disappeared from the stage of history. However, *Shenyi* was the prototype of *Kuzhe* and *Ruqun*.

Before *Shenyi* appeared, clothes for ancient people consisted of the upper part and lower part. *Shenyi* was a kind of loose straight dress covering the upper and lower of the body; however, there remained a boundary line between the upper part and lower part of the body by tailoring halves and sewing together.

Each part of *Shenyi* was given an individual symbolic meaning. In the

祛：指衣袖的袖口，通常是用厚实的织物为材料，起装饰作用，并增强衣袖的耐损程度。其形制要比袖身小，魏晋以前男女都可以使用，主要用于深衣和袍服。

Qu: means cuff. It used to be made of heavy material for the purpose of ornament and standing wear and tear. The size of the cuff was smaller than that of the sleeve. Both men and women could wear *Shenyi* and gown-style dress with *Qu* prior to Wei and Jin Dynasties (220-420).

三重衣：领口很低，能露出里面的衣服，最多的达三层以上，称为"三重衣"。

Triple-layer collar: The collar is low enough to reveal three layers of clothes, hence its name.

交领：形制为长条，下部连带衣襟，穿着后两襟相互叠压，故因穿着后的形态而得名。男女都可以穿用，只要衣领相交，都可称为交领。

Crossed collar: It's long in shape connected with front piece of the gown. Two front pieces of the gown overlap each other which is similar to a crossed collar, hence its name. It's for both men and women.

袂：最初指衣袖的下垂部分，位置在人体胳膊的肘关节处，一般多作弧形，方便手臂的弯曲和伸展。后来逐渐引申为衣袖的统称。

Mei: originally refers to pendent parts of sleeves shown in arc shape on the elbow joint.

右衽：衣襟由左向右掩，以带或纽固定。在中国古代，汉族服饰中多为右衽，而少数民族的多为左衽。

Youren: (right-buttoned) the right forepart of the gown is covered by the wider left forepart and tied by sash or button. In ancient times Han Chinese usually wore coat or gown right-buttoned while the minorities preferred left-buttoned ones.

衣襟：指衣的前幅。

Yijin: refers to front piece of the dress.

腰带：束腰用的巾帛，常以绫、罗、绸等织物为材料，使用时在腰部缠绕数圈。

Sash: It's made of silk for enlacing around waist.

曲裾："裾"是指衣服的大襟，曲裾是指把衣襟环绕形成的衣襟样式。

Quju: *Ju* refers to the full front part of Chinese gown. Winding the front part of a gown forms the pattern of *Quju*.

• 曲裾深衣

曲裾深衣是深衣的一种，前襟被接长一段，穿着时须将其绕至背后。汉代，曲裾深衣是女子服装中非常流行的款式，因为它通身紧窄，衣长曳地，下摆呈喇叭状，行不露足，最能体现女子的婀娜多姿。

Qu Ju Shenyi

It was one kind of *Shenyis*. Elongated front piece had to be wound to the back. In Han Dynasty *Qu Ju Shenyi* was the most popular style among women's clothes. It was a long and tight gown with the trumpet-shape hem. Women looked graceful with their feet veiled.

领，以示规矩，寓意做事要合乎准则；在后背垂直如绳的背线，寓意做人要正直；水平的下摆线，寓意处事要公正、公平。

process of making, firstly, the upper part and lower part were tailed separately and then stitched at the waist, which meant respecting the ancestors and succeeding to old traditions. Round sleeves and square collar meant faying in with a rule. There was a vertical line at the back implying the meaning of being an honest person. Horizontal line on the dress tail implied to be impartial.

- **身穿绕襟深衣的木俑**

  绕襟深衣是在曲裾深衣基础上的变体形式，西汉时期的女子深衣多为此样式，即将衣襟接得极长，穿时在身上缠绕数道，用带子匝结固定，每道花边则显露在外。

  Wooden figurines in *Shenyi* with winding forepart

  Front piece winding *Shenyi* is the variant of *Qu Ju Shen Yi*. Women in Western Han Dynasty (206B.C.-25A.D.) mostly wore this style. They made the front piece of the gown extremely long to wind several times around their body and fixed by sashes revealing every beautiful border.

- **禅衣**

  禅衣是一种单衣，即没有衬里的单层外衣，其外形与深衣相似，有直裾和曲裾两种款式，不论男女均可穿用。

  Danyi

  Danyi is a single layer garment without lining which is similar to *Shenyi*. It includes two styles *Zhiju* and *Quju* suitable for both men and women.

- 杂裾垂髾服

杂裾垂髾服是流行于魏晋南北朝的女性服饰，因在服装上饰以"纤"而得名。所谓"纤"，是指一种衣服下摆部位以丝织物制成的饰物，特点是上宽下尖形如三角，并层层相叠，一般固定在衣服下摆部位；所谓"髾"，是指一种长飘带，从围裳中伸出来，拖至地面，走起路来，如燕飞舞。

Za Ju Chui Shao dress

It was a popular female dress of Wei, Jin, Southern and Northern Dynasties (220-589). Its name comes from *Qian* which is the silk woven ornament on the hem of the gown. It features wide top and tapering at the bottom just like a triangle. It's overlapped and fixed on the hem of the gown. *Shao* is a kind of long ribbon. The extension part of *Shang* is down to the ground wafting with women's swift walking pace.

古人的发式

发式是人类非常重要的装饰。中国古代发式主要有披发、辫发、发髻、剃发等多种形式，其中以发髻的形式最为多样。

在古代的中国，男女都留长发，披发是将所有的头发自然垂下，头发覆面或是将头发由前朝后梳理，头发中间用带子系扎后披搭在背后。至今一些少数民族仍保留着原始的披发习俗。

辫发也是中国古代先民的发式之一，即将头发在颅后聚拢起来，编结成辫子。辫发有多种形式：有从头顶梳起，垂到颅后；或是从颅后的发根梳起，自然下垂；还有把头发编结成辫子，再盘绕在头顶。

发髻在中国古代也较为流行，无论男女，都可将头发绾成髻。发髻可盘绕成各种形状，如螺形的螺髻、椎状的椎髻等。"髻"有"继"的寓意，也有"系"的含义，因此古代女子梳髻象征成年后嫁人生子，来维系家庭的命脉。

Hairstyle of ancient people

Hairstyle, as an important ornament, included *Pifa* (hair falling over the shoulders), braided hair, hair buns and shaved hair etc. Hair buns were more diverse than any other hairstyle.

Both men and women had long hair falling over their shoulders in ancient China. This custom still remained in some ethnic minorities till now. Their hair fell in naturally or combed from the forehead to the back and then tied with a ribbon.

Braided hair was one of the hairstyles of the ancient Chinese ancestors, that is, hairs were gathered at the back of the head and braided into a queue. It also had many variations. One of these styles was to braid on the top of the head and hang on the back. Another style was to braid on the back of the head. Some people braided hair into a queue then wound around the head.

Hair buns were popular among women and men in ancient China. They were usually wound into different shapes such as spiral shape and cone-shape etc.. The pronunciation of bun *(Ji)* is the same as heir *(Ji)* in Chinese meaning bearing children for family so married women wore their hair up in a bun.

• 魏晋时期女子的发式

魏晋南北朝时期女子的发式很有特点，名目繁多，有灵蛇髻、蝉鬓、飞天髻、百花髻、芙蓉归云髻等。此画中的灵蛇髻十分特别且富于变化。灵蛇髻随着梳绾方法的不同，可以创造出多种造型。

Women's hairstyles in Wei and Jin Dynasties

Women's hairstyles in Wei, Jin, Southern and Northern Dynasties (220-589) featured their diversity such shapes of buns as snake, cicada's wings, flying Apsaras, flower-blooming, hibiscus and so on. This illustration shows varied snake-shaped buns. The way to comb and tie make snake-shaped buns diverse.

• 明代女子的发式

明代女子的发式亦有独特之处，明嘉靖以后花样增多，有挑心髻、鹅胆心髻、堕马髻、牡丹髻、盘龙髻等。假髻也是明代女子常用的发式，式样更丰富，一般以铁丝编成，外编以发，戴时罩在发髻上，用簪绾住。假髻有罗汉髻、懒梳头、双飞燕等式样。

Women's hairstyles of Ming Dynasty (1368-1644)

Women's hairstyles of Ming Dynasty were unique. More styles appeared after Emperor Jiajing of Ming Dynasty. They wore round flat buns with jewels, peony-shaped, crouching-dragon-shaped and so on. Fake buns which had various styles used to be a common hairstyle in Ming Dynasty. They normally had wire waved frames and covered by fake hair. Women used one to cover their real bun and tied with hairpins. Fake buns had many styles, such as *Luohan* bun, *Lanshutou* (combing-free) and two-flying-swallow.

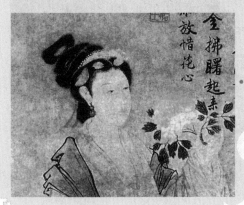

• 唐代女子的发式

中国历代女子发型中，以唐代妇女的发髻式样最为新奇，且名目繁多，如云髻、螺髻、惊鹄髻、峨髻等。

Women's hairstyles of Tang Dynasty (618-907)

Women's hairstyles of Tang Dynasty were the most fancy and diverse. The buns looked like clouds and wings of flying bird.

- 清代满族女子的发式

清代女子的发式分为满汉两种。满族女子的发式变化较多，典型发式有两把头、如意头、架子头、旗髻等。

Manchu women's hairstyles of Qing Dynasty

Women's hairstyles of Qing Dynasty (1644-1911) were fallen into Han and Manchu. Manchu hairstyles were diverse such as double *Batou*, *Ruyi* (an S-shaped ornamental object symbolizing good luck), *Jiazi* bun and Manchu bun.

- 清代汉族女子的发式

汉族女子的发髻在清初时基本上沿用明代发式，清中期开始模仿满族宫廷贵妇的发式，以高髻为尚，名目繁多。有把头发梳成两把的叉子头；有垂一绺头发在颅后，修成两个尖角的燕尾式。清末，梳辫逐渐流行，成为中青年女性的主要发式。

Han women's hairstyles of Qing Dynasty (1644-1911)

In early Qing, Han women's hairstyles were originated from Ming Dynasty. Starting from mid-Qing Dynasty, Han women's hairstyles followed the noble's in the court. They wore their hair in high buns as fashionable style. Some dressed their hair to the shape of *Chazi* (two protruding portions on the top of the head). Some left a tuft of hair on the back of the head trimming it in dovetail shape. Braid, as a main hairstyle of young and middle-aged women, became popular at the end of Qing Dynasty.

> 袍

袍，又称"袍服"，是一种长度通常达膝盖下的服装，战国以后较为常见，男女均可穿着。

战国时的袍服与深衣的最大区别是，袍服为直裾，深衣为曲裾。袍的袖子相对较窄，而深衣多用宽袖。此外，袍的衣摆不像深衣那样宽大。袍与深衣的相同之处是，均为交领右衽，衣裳均连为一体。袍在最开始只作为内衣，需在外面另加罩衣。到了汉代，人们家居时也可以单独穿袍，不需要另加罩衣。东汉时，袍已经成为新娘出嫁时必备的礼服，不分尊卑都可穿着。

袍由内衣变成外衣后，用途更为广泛，上自帝王，下至百官，礼见朝会都可穿用。可以说，袍服是用途最广的古代服装，既可被用作

> Gown-Style Dress

Gown-style dress with length below knees was common to wear after the Warring States Period (476B.C.-221B.C.).

The biggest difference between the gown-style dress of Warring States Period and *Shenyi* was that the full front and back of the gown were straight, whereas *Shenyi* were curvilinear. Compared with sleeves of *Shenyi*, sleeves of the gown were narrow. In addition, *Shenyi* had a looser width of a hem than that of the gown. However, they remained something in common. Both were a one-piece garment with crossed collar and right-buttoned. Initially the gown was taken as the underwear covered with an overcoat. In Han Dynasty (206 B.C.- 220 A.D.), people wore it at home. Gown-style dress had become an essential wedding dress for all brides disregarded social status in Eastern Han Dynasty (25-220).

- **身着曳撒的明宪宗**

曳撒是一种长袍，大襟，长袖，以纱罗、纻丝为料制作。衣身前后形制不同，前身分为上、下两截，后身做成整片。腰部以上与后身相同，腰部以下两侧折有细裥。明初用于官吏及内侍，红色并缀有补子的是有官职之人，青色无补子的是无官职之人。明代晚期，曳撒已逐步演变为士大夫阶层的常服。

Emperor Xianzong of Ming Dynasty in *Yesan*

Yesan was a kind of long robe made of gauze or ramie with large full front and long sleeves. The front and the back of the garment were different in shape. Front side of the dress was divided into upper piece and lower piece while there was an entire piece on the back. Front and back remained the same above the waist. There were pleats on each flank below the waist. *Yesan* was for officials and attendants in early Ming Dynasty. Those wore red with *Buzi* were officials and those wore cyan without *Buzi* were not officials. In late Ming Dynasty *Yesan* became scholar-bureaucrat's casual wear as well as formal dress for banquet.

礼服，也可作为帝王百官以及庶民日常的服装。

袍的主要特征为交领，两襟相交，垂直而下；衣袖宽大，形成圆弧，袖口部分收敛，便于活动；领、袖、襟、裾等部位缀以缘边，起装饰和增强衣服耐磨性的作用。

Ever since the gown was dressed as outerwear, it's popular from the emperor down to all officials. It was used as formal attire and daily dress for the emperor, officials and ordinary people.

The main features of the gown were crossed collar, crossed front pieces falling straight down, loose sleeves in arc shape and tight cuffs, which was for the convenience of body moving. Decorative rims were added on the collar, cuffs, front pieces and back of the gown to make the clothes durable.

- **身着袍服的唐代侍从**
  袍服是隋唐时期男子的普遍服式，一般为圆领右衽，领袖及襟处有缘边修饰，袖有宽窄之分。官员的常服，一般都用织有暗花的料子制作，并用服色区别身份和等级。男子所穿的袍服下端，即膝部通常加有一道横襕（拼缝），即史籍中记载的"袍下加襕"。

Attendants of Tang in gown
Gown as men's popular costume in Tang Dynasty (618-907) usually was right-buttoned with round collar, had decorative border on collar, cuff and foreparts. The gown had two types of sleeves: wide and narrow. Official's casual wear were made of dark cloth with dim flower pattern. The ranks could be distinguished by color. There always was a joint beam at the lower end, near knee part of men's gown, known as "another piece added to the robe" documented in historical records.

旗袍

旗袍是一种特殊的长袍，原指满族的旗人所穿着的袍服，后来专指女子的袍服，名称始见于清朝。辛亥革命后，汉族女子也以穿着旗袍为时尚，并在原来基础上加以改进，发展成为如今名满天下的中国女装代表，被誉为中国的"国服"。 旗袍对于中国女人来说，永远是一种剪不断的情结，虽历尽百年沧桑岁月，其简洁而典雅、华丽而不张扬的魅力依旧不减。

Chi-Pao

Chi-Pao, a special long garment, originally referred to bannermen's gown. Later it referred specially to women's gown-style dress. Its name appeared in Qing Dynasty (1644-1911). After the Revolution of 1911, Chi-Pao became popular among women of Han ethnic group. Based on the original style, Chi-Pao has been evolving. Because Chi-pao is seen as a symbol of China's women dress, it's known as China's "national dress". With its simple, elegant but not blatant design, Chi-Pao has stood the test of time. Thus, it's deeply rooted in Chinese women's mind.

● 身穿白色旗袍的女子
A Woman in white Chi-Pao

> 衫

衫由深衣转变而来，是魏晋时士人的常用服装，并逐渐成为单衣的通称。衫多由轻薄的纱罗制成，袖子宽大，呈垂直型，袖口不收紧，一般采用对襟，衫襟既可用带子系缚相连，又可不系带子，比袍穿用方便，散热性好，适合夏天穿用。

南北朝时期由于受胡服的影响，穿衫的人逐渐减少，晚唐五代时再度流行。唐代士人平时所穿的衫名为"襕衫"，而百姓所穿的衫与士人有所不同，样式较短小，长不过膝，便于劳作，称作"缺胯衫"。除男子外，隋唐和五代的女子也喜欢穿衫，尤其到了夏季，更以穿着宽衫为尚。

宋代沿袭了唐代的服装制度，士人也穿衫。女子则穿纱罗之衫，

> *Shan* (Long Gown)

Shan evolved from *Shenyi,* was worn by scholars of Wei and Jin Dynasties (220-420). Gradually it became the general term of unlined upper clothing. *Shan* was made of light and thin gauze. It had loose sleeves falling straight down. The opposite front pieces were fastened either with sash or without sash. Because this material had good performance for dissipation, it's suitable for summer wear.

Affected by *Hu* dress of Northern and Southern Dynasties (420-589), the number of people who wore *Shan* gradually reduced. However, it came into style again during the late Tang and Five Dynasties. *Shan* worn by scholars of Tang Dynasty (618-907) was called "*Lanshan*", which was different from *Shan* worn by ordinary people. *Shan* for common people was called "*Queku Shan*". It extended above the knees for

但在穿法上趋于拘谨和保守，不像唐代那样袒胸露脯，多加有衬衣。由于衫的衣袖宽大，宋代干脆就称这种女服为"大袖"。到了明朝，衫大多作为女子礼服。随着后世发展，衫开始泛指衣服。

the convenience of labor work. Besides men, women in Sui, Tang and Five Dynasties also liked wearing *Shan*. Loose *Shan* was in fashion in summer.

Following the costume institution of Tang Dynasty, scholars in Song Dynasty (960-1279) wore *Shan*. Women wore gauze *Shan* but it tended to be conventional compared with open style of Tang Dynasty. Women wore *Shan* with underlinen. It had loose sleeves in Song Dynasty, hence its name. *Shan* became female formal attire in Ming Dynasty (1368-1644). With its later development, the term *Shan* meant in general clothes.

- 缺胯衫

缺胯衫是指开衩的短衫，长不过膝，在胯部的前后、两侧各开一衩，衩旁有缘饰，常用白布为料。因为方便穿着与活动，所以多出现在庶民之中，始于唐朝，宋、明时期广为流行。

Queku shan

Queku shan was a short gown above knees. There was trimming edged slit at each flank, front and back of the crotch respectively. It was usually made of white cloth. Its simple style and conveniences for movement made it was worn by plebeian mostly. *Queku shan* appeared in Tang Dynasty (618-907) and became popular in Song and Ming Dynasties (960-1644).

对襟袒胸衫

对襟袒胸衫是魏晋时期文人广泛穿着的一种衫，对襟、大袖，衣长至膝下，衫的对襟常有垂褶，腰间系有腰带，飘洒自如，随心所欲，半披半曳，给人一种放荡不羁、不入世俗的感觉。

Exposing front opening long gown

It was widely worn among scholars in Wei and Jin Dynasties (220-420). It's a kind of front opening gown down to knees with large sleeves and lappet on closing edge. Tied around the waist with sash, dangling freely, the wearer looked as drifting beyond the secular society.

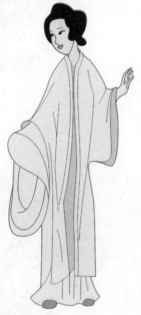

扣身衫子

扣身衫子是一种紧身的衣服，故名。此衫圆领，对襟，长袖，袖身宽松，衣长至膝下，多为妖冶之妇穿用。

Tight-fit gown

It's a tight-fit dress, hence its name. This garment is round collar, front-opening with loose sleeves. Reaching below knees, it was mostly worn by witches.

素罗大袖

素罗大袖是宋代贵妇的流行服饰。衣缘加有花边，两边袖端各接有一端，延长为长袖，接缝处也有花边，此衫多穿在外。这件素罗大袖是非常难得的出土的实物，以单层素罗制成，前身长120厘米，后身长121厘米。

Plain gauze garment with large sleeves

It was popularized among grandee dames in Song Dynasty (960-1279). With laced edge and extended sleeves it was an over gown. The extra pieces on sleeves make it a long-sleeve dress. Even the joint seam was nicely embellished. This is a scarce unearthed garment made of single layer of plain gauze. The front part is 120cm long and back part 121cm long.

云肩

云肩是一种肩背上的衣饰,装饰图案内涵丰富,设计巧妙。云肩平展开来,中心围绕颈部呈放射状,有四方、八方等形态,外观则统一呈圆形,整体上均是外圆内方,象征天圆地方、四方四合、八方吉祥。中国的传统服饰一般均是平面裁剪,而云肩不同,它是立体裁制,因人制宜,非常贴合女子圆柔的肩部轮廓。云肩穿在身上,层层叠叠,有的四周垂下吊穗,配合所绣的四季花果和渔樵耕读等劳动生活场景,象征着人与自然的融合。

云肩始自隋朝,至明清时期普及。明清贵妇礼服上,云肩是常见的装饰。相传清代慈禧太后的一件云肩竟由3500颗又大又圆的珍珠穿织而成,灿烂夺目,价值连城。清代的江南女子时兴梳一种低垂的发髻,还喜欢在发髻上抹发油,显得头发乌黑有光泽。为了避免衣领和肩部被发髻油腻沾污,也流行披云肩。

• 清代身披云肩的女子
Womens wearing cappa in Qing Dynasty (1644-1911)

Cappa

As an ornament on shoulders, it had an ingenious and artful design. Spreading a cappa on shoulders, it presented the patterns radiating from the neck to four directions or different directions. Generally, it's designed as round outside and square inside to symbolize hemispherical dome, round heaven and square earth, endowing an auspicious meaning. China's traditional clothing mostly adopted flat cutting but cappa was different, which used three-dimensional cutting. It suited measures to different ladies and catered for gentle outline of their shoulders. Shoulders were covered with cappa, layer upon layer, hanging panicles to match the embroidered patterns. Patterns on cappa were associated with everyday life such as flowers, fruits of four seasons, fishing and farming symbolizing the integration of man and nature.

Cappa first appeared in Sui Dynasty (581-618). It didn't gain popularity until Ming and Qing Dynasties (1368-1911). It was a common part on the formal attire for grande dames in Ming and Qing Dynasties. It's said that Empress Dowager *Cixi* had a dazzling and priceless cappa which was woven by 3,500 large and round pearls. Women in southern China wore their hair in low buns, which was fashionable in Qing Dynasty. Women used to oil their hair to have black sheen, thus they wore the cappa to avoid smearing their collar and shoulder.

• 各式云肩
Different patterns of cappas

手帕与团扇

手帕又称"手绢""汗巾",是古人随身携带,用于擦手、脸的方形织物。东汉时期,人们就开始使用手帕。唐朝是中国服饰文化发展的鼎盛时期,"手帕"这一名称此时被正式推出。随着时代的变迁和人类文明的发展,手帕除原有功能之外,还具有浓郁的审美情趣。明朝中叶以后,人们常把一些女子结成深交称为"手帕姐妹"。女子时常将手帕别在腰际或拿在手中,既方便随时擦拭,又极为美观。

团扇,又称"扇子",是古代女子的随身之物,也是一种装饰品。西汉时期,团扇的使用就已较普遍了,造型以圆似明月居多,其他形制的有梅花形、海棠形、马蹄形、六角形、八角形、瓜棱形、蕉叶形、梧桐叶形等。隋唐时期,由于造纸技术发展较快,纸面团扇也渐渐流行起来。唐末宋初,刺绣工艺已非常精妙,一些花鸟、鱼虫、人物、山水的刺绣开始出现在团扇的扇面上。到了宋代,社会上书扇、画扇、卖扇、藏扇、玩扇的风气很盛,扇子作为一种艺术商品广为流通。清代,扇文化和工艺进入巅峰。扇子主要材料有竹、木、纸、象牙、玳瑁、翡翠等,飞禽翎毛、麦秆、蒲草等也能编制成形态各异的日用工艺扇。其造型优美,构造精致,扇面幅不盈尺,经能工巧匠镂、雕、烫、钻,或名人挥毫题诗作画,或刺绣山水花鸟、人物故事,团扇艺术身价倍增。古代女子都喜欢手执团扇。清代宫廷里最流行的搭配是这样的:浓妆淡抹的女子,身穿一袭丝绸锦缎制成的宽袖大摆的旗装,轻执一柄玲珑剔透的团扇,纤纤玉手上戴一只金镯或翠镯。

Handkerchief and Moon-shaped fan

Handkerchief, a square-shaped textile also known as *Shoujuan* or *Han Jin* in Chinese, used to wipe hands and face. In Eastern Han Dynasty (25-220), people began to use handkerchief. The development of China's costume entered a period of great prosperity in Tang Dynasty (618-907) when the name of handkerchief was officially used. With the changing of times and evolution of human's civilization, handkerchief showed a strong

- 宋代握手帕的女子
  Woman in Song Dynasty with a handkerchief

- 清代手执团扇的女子
  Woman in Qing Dynasty with a moon-shaped fan

aesthetic taste in addition to its basic function. Chinese people often described female close friends as "handkerchief sisters" after the mid-Ming Dynasty. Women usually pined handkerchiefs on their waist not only for the convenience of wiping hands but also for decoration.

  Ancient Chinese women brought with a fan for decoration. It's commonly used in Western Han Dynasty (206B.C.-25A.D.). They used different shapes of hand fans such as round fan which accounted for the majority, fans in plum blossom shape, fans in Chinese crabapple flower shape, fans in U-shape, fans in hexagon, fans in octagon, prismatic-shaped fans, fans in banana leaf shape and fans in leaf of phoenix tree shape etc. Owing to the rapid development of paper-making technology, paper made moon-shaped fans gradually became popular in Sui and Tang Dynasties (581-907). Embroidery was very subtle at the end of Song and the beginning of Tang Dynasties. Such designs as flowers, insects, fish, figures and landscape were embroidered on fans. Fans became the artistic commodity in Song Dynasty (960-1279), calligraphy on fans, fan sales and fan collection prevailing at that time. Culture related to fans and the making process of fans reached the summit in Qing Dynasty (1644-1911). Major materials included bamboo, paper, wood, ivory, hawksbill and emerald. Different kinds of artistic fans could be made of feathers, wheat-straw, stem and leaf of cattail. The elaborate and elegant fans were made in the process of carving, engraving, pressing and drilling. Because fans were embroidered famous people's inscription and painting, beautiful landscape, birds, flowers, and people's story, the value of fans redoubled. When women in ancient times wore skirts, they preferred bringing with moon-shaped fans. Fashionable match in Qing court could be seen in some of paintings. Holding an exquisite round fan, she wore a loose-sleeved Mongolian dress made of damask and silk. An emerald bracelet or a gold bracelet was around her slim wrist.

> 襦

襦是一种短衣，一般下身配裙穿着。东汉前，男女都可以穿襦，既可当衬衣，又可当外衣；东汉以后则多为女子穿用。

汉魏时期，襦一般采用大襟，有长短之分，长者至膝部，短者至腰间；衣袖有宽窄两种样式；有单夹之分。隋唐时期，襦的式样有所变化，

> *Ru* (Short Jacket)

Ru was a kind of short jackets to match the skirt on the lower part of the body. It was for men and women before Eastern Han Dynasty (25-220) and could be used as shirt as well as coat. After that it was more specific for women.

During Wei and Han Dynasties, *Ru* was a side-opening jacket. It had short style and long style; the length of long style reached knees; the length

- **身着襦裙的仕女**
此图描绘了几名五代时期的奏乐女子，皆梳高髻，上穿襦，下穿长裙，色彩比较淡雅清新。
Beauties wearing *Ruqun*
This illustration portrays five girls in Five Dynasties (907-960) who play the musical instruments. Wearing their hair in high buns, they wear *Ru* on the upper part of the body and long skirt on the lower part of the body in plain colors.

- **身着襦裙的唐代女子**

  此图描绘了一群唐代女子捣练、络线、熨烫及缝衣时的情景。她们着装为典型的唐代女服样式，梳高发髻，穿窄袖短襦，下着长裙，裙腰高系，裙长曳地。

  Women of Tang Dynasty (618-907) in *Ruqun*

  This illustration presents the scene of a group of women in Tang Dynasty selecting, winding strings, ironing and sewing the clothes. They wore dress with typical features of Tang Dynasty. Wearing their hair in high buns, they wear short *Ru* with narrow sleeves. Their long skirts extend to the ground with sash tied high on the waist.

更多采用对襟，不用纽带，以窄袖为主，袖长至手腕或超过手腕，多与裙配套穿用，下束于裙内，因此有"襦裙"之称。到了宋代，由于褙子的出现，穿襦者一度减少，但士庶女子还

of the short style reached waist. There were two kinds of sleeves. One was narrow. The other was loose. *Ru* had a single layer style and a lined style. The styles of *Ru* varied in Sui and Tang Dynasties(589-907). It's a kind of

常穿用。元代，襦又重新流行起来。直到清代中期，由于袄的流行，襦才逐渐消失。

opposite front pieces garment without sash. Its narrow sleeves reached around wrist. It's also known as *"Ruqun"*. By Song Dynasty (960-1279), with the emergence of *Beizi*, the number of people who wore *Ru* became less and less but it still retained. *Ru* came into fashion again in Yuan Dynasty (1271-1368). When a kind of cotton-padded jacket popularized in mid-Qing Dynasty, *Ru* gradually disappeared.

- 明代女子襦裙的样式

明代女子一般上身穿襦，下身穿裙，襦的样式为交领、长袖短衣，下穿裙，裙内加穿膝裤。腰上往往挂一根以丝带编成的"宫绦"，有的还在宫绦上串块玉佩，用以压住裙幅，使其不致散开影响美观。

The pattern of women's *Ruqun* in Ming Dynasty(1368-1644)

Women in Ming Dynasty wore *Ru* on the upper part of the body and wore skirt on the lower part of the body. *Ru* is a short jacket with long sleeves and a crossed collar. People wear a skirt on the lower part of the body and *Xi Ku*(trouser legs). They mostly tie a *"Gongtao"* woven by ribbons. Some put a jade pendant in *Gongtao* to pin down the folds on the skirt.

荷包与香囊

荷包是古人用来盛放散碎银子、票据、印章、手帕、针线等零碎物品的小型佩囊。荷包多皮制，也有丝织的，既可提在手上，又可背在肩上，后来逐渐演变成佩挂在腰际的袋囊。荷包多用丝织物做成，质地柔软，有圆形、椭圆形、方形、长方形、桃形、如意形等各种形状，绣有花草、鸟兽鱼虫、山水、人物、传统吉祥图案、祈福祝祷之语、诗词文字等纹饰，精巧别致，极富装饰性。"荷包"一词最早出现在宋代，发展到清代时，荷包的造型已丰富多样：上小下大，中间收缩，像一个葫芦的，称为"葫芦荷包"；上大下小，形似鸡心的，俗称"鸡心荷包"……

香囊是用布、绸缎等布料缝制的小袋，宫廷及富贵人家的香囊则由金银制造，专门用来贮放香料，佩戴在身上或悬挂于床帐内。早在先秦时期，就有佩挂香囊的风俗。中国古代端午节时，儿童多佩戴香囊，不仅辟邪驱瘟，还起装饰作用。香囊除了作为佩饰，还是青年男女的定情信物。女子常将佩戴的香囊相赠男方，以表心迹，留作纪念。民间的香囊多用素罗制成，绣鸳鸯、莲花等寓意吉祥的图案。明清时期，香囊更成为青年男女间流行的传情之物，属于非常私密的闺房用品，通常是年轻的已婚之人佩戴。

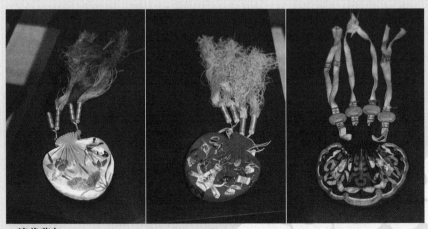

- 清代荷包
  Pouches of Qing Dynasty（1644-1911）

Pouch and perfume sachet

Pouch was a small bag for carrying pocket money, receipts, seals, handkerchiefs, odds and ends in ancient times. Pouch was mostly leather- made and it was also silk-woven. It's carried either by hand or by shoulder. Later it evolved into a pouch hanging on the waist. Pouch was in various shapes such as round, square, ellipse, rectangle, peachlike shape and *Ruyi* (S shaped ornamental object, a symbol of good luck). Pouch was mostly made of lustring so it felt soft. Pouch was so extremely delicate that such various patterns as flowers, grass, birds, beasts, fish, insects, landscape, figures, traditional auspicious patterns, blessing words and poems were embroidered on it. The word "pouch" first appeared in Song Dynasty. Pouch had various shapes during Qing Dynasty(1644-1911). Cucurbit-shaped pouch was large on upper and small at the bottom constringing in the middle. Heart-shaped pouch had bigger top and smaller bottom.

Perfume sachet was a small bag made of cloth, silk and satin. Wealthy people and people in the court using gold and silver made sachet for perfume storage hung on the bed curtain and on the body. As early as pre-Qin period, there was the custom of wearing sachets. In ancient China, children wore sachets on Dragon Boat Festival to drive away evil and plague. It's used for decoration as well. A woman gave her sachet to a young man as a token of love. Sachets were made of plain gauze among common people. They were embroidered mandarin duck, lotus flower and other auspicious patterns. In Ming and Qing Dynasty（1368-1911）, sachets prevailed among women and men to express their love, thus sachets became very intimate items which were worn by the young married.

• 清代金丝香囊
Spun gold sachet of Qing Dynasty

> 袄

袄是从襦演变而来的一种短衣，比襦长，比袍短，衣长多至人的胯部。袄多用质地厚实的织物制成，保暖性强，用作秋冬服装。缀有衬里的袄，称作"夹袄"；冬季用时，中间纳絮棉花，俗称"棉袄"。袄的基本款式是以大襟窄袖

> *Ao* (Short Jacket)

Ao evolved from a short jacket, longer than *Ru*, shorter than gown, reaching to crotch. It was made of heavy textile to keep warm in winter. With the lining inside, it was known as "lined jacket". In winter, it was padded with cotton known as "cotton-padded jacket". The basic style of *Ao* was a full front garment with either narrow long sleeves or narrow short sleeves. Another style had opposite front pieces varied in length of sleeves.

Ao appeared around Wei, Jin, Southern and Northern Dynasties. It was firstly worn by the northern ethnic minorities. After Sui and Tang Dynasties, when it was spread to the Central Plains, it became popular among

• 清代的夹袄
Lined jacket of Qing Dynasty

为主，也有对襟，袖有短袖、长袖之分。

袄大约出现在魏晋南北朝时期，最初为北方少数民族穿着，隋唐后传入中原，广为流行，不分男女，除重大朝会，平时皆可穿着。到了明清，袄更成为士、庶女子的主要便服，多与裙搭配穿用。

women and men. People wore it in daily life with exception on the occasion of important court meetings. By Ming and Qing Dynasties, *Ao*, as an informal dress, was worn by scholars and common women to match skirt.

黄道婆与棉布

在棉花普遍种植以前，古人的衣料一般为丝或麻。在中国，棉花最初只在南方少数民族地区种植，时称"白叠"。到了唐代，中原地区开始使用棉花，并引进其种植技术。元代初期，棉花在江南普遍种植。宋末元初，一个叫黄道婆的女子漂流到海南岛，从当地人那里学会了棉纺技术。后来，她回到中原，革新了纺车，向人们传授棉纺织技术，使棉纺织业在汉族地区迅速发展起来。从此，棉布取代了麻布的地位，成为中国人主要的衣料之一。

Huang Daopo and cotton cloth

Before planting the cotton, ancient people used silk and flax as main materials for clothes. Cotton was planted in southern China where the ethnic minorities lived, named *Baidie* at that time. People in the Central Plains started to introduce cultivation technology and use cotton in Tang Dynasty (618-907). At the beginning of Yuan Dynasty, cotton was widely grown in southern China. At the end of the Song Dynasty and the beginning of Yuan Dynasty, a lady named Huang Daopo drifted to Hainan island to learn spinning technology from local people. When she returned to the Central Plains, she updated the spinning wheel and imparted people to spinning technology, which pushed the rapid development of cotton spinning technology in Han nationality. Thereafter flax was replaced by cotton and became one of the major materials in China.

暖耳

暖耳是中国古代冬季御寒、保护耳朵的服饰。暖耳最早见于唐代,明代时曾列入官服制度,平民百姓禁用。到了清代,暖耳不再禁用。暖耳一般用狐皮制作,有些用棉布制作,内絮棉花。

Ear cover

Ear cover, as a decoration for ears, used in winter to keep warm. It appeared in Tang Dynasty (618-907). In Ming Dynasty (1368-1644), it was listed into the official uniform institution and common people were not allowed to wear. By Qing Dynasty (1644-1911), this restriction was rescinded. Some ear covers were made of fox fur. Others were made of cotton cloth padded with cotton.

• 清代暖耳
Ear cover of Qing Dynasty

> 褂

褂在明代时指罩在外面的长衣，清代时则指罩在袍服之外的外服，男女均可穿着。褂有两种样式：一种衣长至膝，用于行礼者，称作"长褂"，若缀有补子，又称"补褂"；一种衣长仅至胯部，用于出行，称为"马褂"。马褂的式样为圆领或立领，大襟或对襟，褂长至腰，下摆开衩，袖口部分大多齐平。

> *Gua* (Jacket)

Gua of Ming Dynasty referred to a long garment over other clothing. In Qing Dynasty, it referred to a garment over the gown-style dress. Both men and women wore it. There were two types of *Gua*. One for rituals which extended to knees was named "Long jacket". If there's *Buzi* attached, it's also called "*Bugua*". The other was called "Ridding jacket" with the length reaching crotch for convenience. There were additional styles of "ridding jacket" - stand collar or round collar, garment with full front or opposite front pieces, a slit on the lower hem and even cuffs.

缂丝

缂丝是古代一种特殊的织造工艺，盛于宋代，清乾隆时期也很流行，是用较坚挺的半熟丝作经线，柔软的全熟丝作纬线，织成各种花纹。由于丝线材质的差别使得花纹连接处呈现出明显的刀刻般的断痕，成品图案效果犹如细细雕琢篆刻出来一般玲珑浮凸、清晰生动，故得其名。缂丝工艺除用于服饰外，还用来织造屏风类观赏品。

Gesi Tapestry

Gesi Tapestry, contexture technology in ancient China, flourished in Song Dynasty and prevailed in Emperor Qianlong of Qing Dynasty. Various patterns were woven by firm semi-cuit silk as warp and soft cuit silk as weft. Because of difference in silk thread, the effect of carving cracks was presented on the connecting points of the patterns. The finished part was as vivid as a living figure just like an exquisite relief in fine engraving. Besides costumes making, tapestry process was also used for weaving folding screens.

• 清代缂丝马褂
Ridding jacket of Qing Dynasty made of gesi tapestry

御赐"黄马褂"

　　黄马褂是清代皇帝赐予侍臣及有功勋人员的短衣，罩在袍之外，素色无纹样，不施缘饰，用明黄色绸缎制作。黄马褂分为两类：一类为大臣、侍卫跟随皇帝出行时穿着，以彰显皇家威仪；另一类为皇帝奖赐给军功人员的服装。御赐黄马褂是清代最高的荣誉，一般只有立过功勋的高级军将或统兵文官才能得到。

"Yellow jacket" bestowed by emperor

Qing emperor bestowed Yellow jacket to courtiers and meritorious personnel. It was a short plain jacket without pattern and decorative rim on it. It's made of silk and satin in bright color and worn over the gown. Yellow jacket was divided into two categories: one was the ministers and the guards followed the emperor wearing it to show majesty of the royalty; the other was granted to meritorious military personnel. Yellow jacket was regarded as the highest honor in Qing Dynasty. Only high officials and generals who had meritorious deeds had the honor to be bestowed.

● **身着黄马褂的清代官员**
The official of Qing Dynasty in Yellow jacket

> 裳

裳，又称"围裳"，商周时期，裳是遮蔽下体的服装统称，无论男女、尊卑都可以穿着。

裳的形制与后世的裙十分相似，但区别在于裙多被做成一片，穿时由前向后围；而裳则分成前后两片，一片蔽前，一片蔽后，前片由三幅布组成，后片由四幅布组

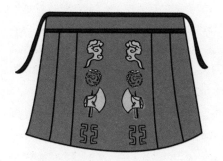

• 西周时期的裳
Shang of Western Zhou Dynasty (1046 B.C.-771 B.C.)

> *Shang* (Dress on the Lower Part of the Body)

Shang, as a general designation of dress covering lower part of the body, could be worn by anyone disregarded social status.

The shape of *Shang* was much similar to the later skirt but there was difference in width of cloth. Skirt was made in whole piece, which was worn from the front around to the back while *Shang* had two pieces with one covering the front and the other covering the back. Three widths of cloth were jointed to make the front piece. Back piece was combined with four widths of cloth. There were folds on the top of *Shang*. The number of folds was determined by the size of the wearer's waist.

When wearing *Shang*, people tied the sash around the waist. Pants emerged in Shang and Zhou Dynasties (1600B.C.-256B.C.) which were totally different

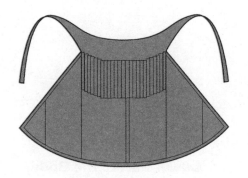

- 雨裳

雨裳是一种用毡、皮、羽纱或油绸等料做成的围裳，下雨下雪时穿戴，不怕潮湿。一般穿在雨衣之内，长至足面。清代广为流行，皇帝与文武百官均可穿用，仅以颜色区别等级。

Yu Shang

Yu Shang is made of felt, leather, camlet and oiled silk. People wear it around the body in rainy day to keep away from wetting. It's down to feet worn inside the raincoat. Widely popular in Qing Dynasty (1644-1911), both the emperor and officials could wear it using different colors to distinguish ranks.

成。裳的上端有折裥，裥的多少视穿裳者的腰身粗细而定。

穿裳时还要配腰带，系在腰上。商周时期已出现了裤，但其仅有裤管，没有裤裆，穿着时套在腿上，外面还要着裳，用来遮蔽隐私部位。因裳紧贴下身，被视为卑亵之物，所以裳的实际作用是用来遮羞的。裳的前后分制，两侧有缝隙，可以开合，故穿着时举止上必须加倍注意。中国在汉代以前最常见的坐姿是跪坐，正与此有关。

汉代以后，出现了有裆之裤，裳的前后两片被连在一起形成了裙，并逐渐取代了裳，与襦、袄等

from the present pants. Pants at that time were worn around legs without crotch. *Shang* was worn over the pants to veil the private part of the body. Because *Shang* was tightly close to the private part of the body, it's seen as a profane thing. There was a seam at each flank of *Shang* which could be open. Therefore, People had to behave themselves properly. That accounted for the reason why people had knees sitting prior to Han Dynasty (206B.C.-220A.D.).

After Han Dynasty, crotch appeared on the pants. Skirt formed after front piece and back piece of *Shang* were connected. Gradually it replaced *Shang* and matched *Ru* and *Ao* to become

上衣相搭配，成为女子的主要服装。而男子的便服，则多为袍服，不过裳并未彻底消失，而是与上衣相配合使用作为礼服的主要形制，在举行典礼时使用。裳的形制和颜色是区分身份等级的重要标志。

the major costumes for women. Men's informal dress was mainly the gown-style dress but *Shang* didn't disappear completely. Matching the upper jacket *Shang* became the formal attire worn on the occasions of ceremony. People's social status and ranks were distinguished by the shape and color of *Shang*.

古代女子的妆容
Makeup in ancient China

中国古代人面部的妆饰，一般施用于女子。早在商周时期，古人就已经开始在面部进行妆饰了，至隋唐五代时尤为盛行。面部妆饰常见的手法包括画眉、敷妆粉、涂脂、面靥、额黄、斜红、花钿、点唇等名目。使用的材料包括妆粉、胭脂、石黛等。妆粉是用米粉和铅粉制作的改善面部肤色的化妆品；石黛是一种用来画眉的黑色矿物质；胭脂则是从一种叫"焉支"的植物中提取的红色液体，并将其与动物膏脂混合制成。

As early as Shang and Zhou Dynasties (1600B.C.-256B.C.), ancient women began to do facial makeup. It prevailed in Sui, Tang and the Five Dynasties (589-960). Facial makeup included brows painting, putting makeup powder on cheeks and adding rouge to lips etc. The materials included makeup powder, rouge and *Shi Dai*. Facial powder was made of rice flour and lead powder to improve facial color. *Shi Dai*, a kind of black mineral was used to paint eyebrows. The red color of rouge was extracted from a plant and blended with animal fatty cream.

敷妆粉：中国历代不同时期的审美习惯使面妆的种类异彩纷呈，如汉代的白妆，魏时的斜红妆，唐代艳丽桃花妆、酒晕妆，宋代素雅的薄妆等。然而各个时代的化妆习俗虽然不尽相同，但基本上都有以肤白为美的特点。

Makeup powder: Facial makeup varied from Chinese aesthetic in different dynasties. Even if there were different makeups such as White-style makeup of Han Dynasty (206B.C.-220A.D.), *Xie Hong* makeup of Wei Dynasty, heavy makeup, drunk-style makeup of Tang Dynasty (618-907), and light makeup of Song Dynasty (960-1279), they featured fair skin.

● 酒晕妆

酒晕妆因两颊胭脂浓艳，好像酒后酡红的两颊而得名。

Drunk-style makeup

Putting the bright-colored rouge on the cheek looks like a lady in drunk.

● 薄妆

薄妆是流行于宋元时期的一种淡妆。当时的女子因受礼教的束缚，大多摈弃了唐代的那种浓艳的妆容，转而崇尚素雅、浅淡的妆饰，只在脸上薄薄施以浅淡的朱粉，透出微红。

Light makeup

Light makeup was popular in Song and Yuan Dynasties (960-1368). When women were shackled with feudal ethics and rites, most of them discarded extravagant makeup style and turned to light makeup. They only put light rouge on cheeks revealing reddish.

● 晓霞妆

晓霞妆是一种将胭脂描绘在鬓眉之间的面妆，形状多样，有的是月牙状，有的如残破滴血状，有的则作卷曲花纹状。

***Xiao Xia* makeup**

Xiao Xia makeup is to put rouge between eyebrows and temples varied in shapes, such as crescent shape and curl type.

画眉：画眉是指以石黛等材料涂染眉毛，以修饰和改变眉形的化妆术。早在战国时期，画眉之风就已经出现，及至秦代已相当普及。到了两汉时期，上承先秦列国之俗，下开魏晋隋唐之风，画眉成为风尚。至隋朝时，因隋炀帝喜好长眉，从波斯进口大批昂贵的螺黛，日供五斛仍不能满足后宫的需要，迎来画眉史上的一个高潮。到了唐代，女子画眉已成习尚。继唐代之后，画眉之风依然广为流行，直到近代，仍为广大女子所喜爱。依据各个时期的不同审美，各朝的画眉喜好也各不相同。秦汉时期崇尚长眉，直到隋代，这种纤细修长的眉式依然深受女子喜爱。唐代女子眉型偏好浓厚。宋代眉型则趋清秀。元代后妃眉型一律都为一字眉，这也是蒙古贵妇的特有妆容。及至明清时期，女子崇尚秀美，眉型大都纤细弯曲，长短深浅等变化日益减少。

Eyebrows painting: using the graphite to paint eyebrows so as to change the shape of brows. Brows painting appeared as early as Warring States Period. It became popular in Qin Dynasty (221 B.C.-206 B.C.). Following the old custom of pre-Qin and carrying out the new custom for Wei, Jin, Sui and Tang Dynasties, eyebrows painting prevailed. In Sui Dynasty (589-618), emperor Yang of Sui preferred brows painting. A large quantity of expensive *Luo Dai* (a kind of brows painting material in dark green) imported from Persia. Even five *hu* (units of capacity) couldn't meet the demand of the imperial harem. It ushered in a climax in history. By Tang Dynasty (618-907), brows painting had become a common practice. It remained to be popular after Tang Dynasty. It's loved by women of modern times. Based on the different aesthetic tastes, brows painting varied in different dynasties. People in Qin and Han Dynasties preferred slim eyebrows. It's also loved by people in Sui Dynasty. Women of Tang Dynasty were in favor of heavy makeup. The shape of eye brows tended to be slim. All empresses of Yuan Dynasty (1271-1368) had straight-line-shaped brows, which was unique makeup for Mongolian grandee dames. Women in Ming and Qing Dynasties (1368-1911) liked slim and bending brows with little variation in length and color.

蛾眉

蛾眉是一种弯曲而细长的画眉样式，秦汉时期就已经出现，因为其形状和蚕蛾的触须相似而得名。这种眉式历代都很流行，后来还演变成对美女的代称。

Eyebrows painted in a moth's antenna shape

It's an eyebrows painting style. Painted eyebrows are slim and bending like moth's antenna, hence its name. It appeared in Qin and Han Dynasties and became popular in following dynasties. Later it evolved into a synonym for beauty.

柳叶眉

柳叶眉属于细长的眉形样式，因其形状为中间宽阔，两头尖细，类似柳叶而得名。这种眉式历代女性均有采用，隋唐五代时期最为盛行。

Willow-leave-shaped eyebrows

Painted eyebrows are wide in the middle tapering on each end and looks like willow leaves, hence its name. This shape of eyebrows could be seen in every dynasty. It was popular in Sui, Tang and the Five Dynasties (589-960).

愁眉

愁眉是流行于东汉后期的一种女子眉式，形状细长而弯曲，两梢下垂，形似哭泣时的样子。

Knitted brows

It became popular in the late Eastern Han Dynasty. Women paint bending and slim brows with each end hanging down. It looks like weeping with frown forehead.

分梢眉

分梢眉是流行于唐代的一种女子眉式，形状为内端尖细而外端宽阔且向上翘，并且在眉梢画出分梢状。

Branch brows

It prevailed in Tang Dynasty (618-907). The tips of the brows are painted branches with inner ends tapering and wide outer ends tilting.

阔眉

阔眉是一种眉形宽阔的画眉样式,出现于西汉时期,盛行于唐代,因其眉形比正常的眉毛宽出数倍而得名。

Broad brows

Brows are painted in broad shape. It appeared in Western Han Dynasty and prevailed in Tang Dynasty. The width of broad brows is several times wider than regular brows, hence its name.

桂叶眉

桂叶眉是流行于中晚唐时期的一种女子眉式,因其形状短而宽阔,像桂树的叶子而得名。

Laurel-leaf-shaped brows

It became popular in the middle and late Tang Dynasty. It's short and broad shape which likes a laurel leaf.

一字宫眉

一字宫眉是流行于元代后宫的一种女子眉型,其形状直而长呈"一"字状。

Straight-line-shaped eyebrows

It was popular in the imperial harem of Yuan Dynasty (1271-1368). Brows present a straight and horizontal line.

点唇：点唇是指以唇脂点染嘴唇，以修饰和改变唇形的化妆术。早在先秦时期，就已经出现女子崇尚点唇的现象。汉代时，点唇的习俗已经形成，且历代盛行不衰。根据点唇的手法、形状和色彩等不同，出现了名目繁多的种类，尤其以唐代的唇样最为丰富多彩。中国女子的点唇样式，一般以娇小浓艳为尚，最理想、最美观的嘴型就是像樱桃那般娇小、鲜艳，俗称"樱桃小口"。

Adding rouge to lips was a makeup to embellish and change the shape of lips. As early as Pre-Qin Dynasty, there was a trend to embellish the beautiful lips. Adding rouge to lips became the custom in Han Dynasty (206B.C.-220A.D.). It prevailed from one dynasty to another dynasty. There were different techniques, shapes and colors of adding rouge to lips. Many different kinds emerged especially in Tang Dynasty. The prevalence of little and bright-colored lips became the basic pattern of lip makeup. Thus the ideal and beautiful outline of lips looked like a cherry which was little and bright-colored, usually known as "cherry-shaped lips".

花钿：花钿是古代女子的一种时尚妆饰，在唐朝时最为风行，有各种形状，颜色以红色、黄色、绿色为主。花钿的做工精细，制作材料多样，一般用胭脂涂画，或由金箔、珍珠、云母、螺钿、彩纸、丝绸、鱼骨、昆虫翅膀、翠鸟羽毛等剪制而成。以金箔片做的花钿，金光灿烂，有的薄如蝉翼，或是做成蝉形，称为"金蝉"。花钿一般点在额头眉间，也可贴在眉宇之间、鬓角、两颊、嘴角、发髻上。

• 妆点有花钿的唐代仕女
Beauties in Tang Dynasty are ornamented with *Hua Dian*.

Hua Dian, ancient women's fashionable ornaments, prevailed in Tang Dynasty (618-907). It had red, yellow and green colors varied in shapes. *Hua Dian* was made of diverse materials with fine workmanship. It's painted by rouge and cut from the material of gold foil, pearl, mica, mother-of-pearl inlay, colored paper, silk, fishbone, insect wing and kingfisher feather etc. *Hua Dian* which was made of gold foil looked as thin as shining cicada wings. Some were made in the shape of cicada known as "golden cicada". *Hua Dian* was mostly worn on the forehead between the eyebrows. It could be pasted on cheeks, corners of the mouth, temples and hair buns.

- 花钿的各种样式
  Various types of *Hua Dian*

> 裙

裙是中国古代女子的主要下装，由裳演变而来。裙是由多幅布帛连缀在一起组成的，这也是裙有别于裳之处。

古人穿裙之俗始于汉代，并逐渐取代下裳成为女性单独使用的服饰。那时的女子穿裙，上身要搭配襦或袄。魏晋以后，裙子的样式不断增多，色彩搭配越来越丰富，纹饰也日益增多。宽衣广袖、长裙曳地是贵族服装的主要特点。这一时期，裙不限于女性穿用，也是贵族男性的常见装束，如"裙屐少年"成为富家子弟的代称。

南北朝以后，裙子才逐渐成为女性的专属服装。两晋十六国时期，一种名为"间色裙"的裙子广为流行。它由若干条上宽下窄的布

> Skirt

Skirt has been the ancient women's main garment for lower part of the body. It evolved from *Shang*. Skirt was made of several connected widths of cloth which differed from *Shang*.

The custom of wearing skirt started in Han Dynasty (206B.C.-220A.D.). It gradually replaced *Shang* for women's exclusive use. Women wore skirts to match *Ao* or *Ru* at that time. After Wei and Jin Dynasties (220-420), the styles and the colors of skirts became diverse. Loose garment and long skirt down to the ground became the main feature. During that time, there's no restriction for women to wear skirt. It's a common dress for noble men, thus the idiom *"Qun Ji Shao Nian"* (man with skirt and wooden shoes) was a synonym of a man coming from a wealthy family.

Skirts gradually became women's

料拼接制成，使女性身材显得修长。隋代，裙的式样承袭了南北朝时的风格，间色裙也仍为女性普遍穿用。

唐代的裙子以宽博为尚，裙幅有六幅、七幅、八幅、十二幅不等，裙摆长度也明显增加。女子在穿裙时，多将裙腰束在胸部，甚至

exclusive clothing after Southern and Northern Dynasties. A kind of "Skirt with pieces together" sewing a certain number of cloth pieces in the same width together prevailed in Jin, Southern and Northern Dynasties as well as Sixteen Kingdoms Period. Women looked slender on it. This style inherited from Southern and Northern Dynasties (420-589). "Skirt with pieces together" was commonly worn by women.

In Tang Dynasty (618-907) women's skirts varied in widths including six widths, seven widths, eight widths and twelve widths. The hem of skirt was obviously lengthened. When women wore the skirts, they usually bound the waist of skirt around their chests beneath the oxter. In this way, it would make skirts look long and slim. It remained the style of Sui Dynasty (589-618). Considering the length of the skirt down to the ground wasn't fit for labor work and

- 间色裙

间色裙流行于两晋时期，是指用两种以上颜色的布条间隔缝制而成的裙子。在制作时，整条裙子被裁剪成数条，几种颜色相互间隔，交映成趣。

Skirt with pieces together

It was popular during Wei and Jin Dynasties (220-420). The skirt is sewn by strips of cloth in above two colors. The whole skirt is sewn in different colored strips.

到腋下，以使裙子显得修长，这与隋代装饰风格一致。由于曳地宽裙不便于劳作，而且在用料上极其浪费，以致朝廷下禁令加以限制："女子裙不过五幅，曳地不过三寸。"宋代沿袭了唐代的裙装风格，裙身仍然宽博，裙幅以多为尚，褶皱很多。

it was a waste of cloth, this style of dress was restricted by the court. As the order of the court the width of women's skirt shall not exceed five widths, its length no more than three *Cun* (Chinese size unit). It was followed in Song Dynasty, that is, wide skirt, with many widths and lots of folds.

In Liao, Jin and Yuan Dynasties, women of Han were dressed basically following the style of Song Dynasty, while ethnic minorities still retained their characteristics. Taking the Khitan and Jurchen ethnic minorities of Liao and Jin Dynasties as examples, they were mostly in dark purple with a whole embroidered flower and *Tuanshan* (a kind of gown) was worn outside.

Dress in Ming Dynasty retained

• **罗裙**

罗裙是以罗为面料制成的裙子，主要流行于唐代。画中的唐代仕女头梳高髻，身穿圆领袒胸罗衫，披红色帔帛，下着双色相间的罗裙，双臂舒展，正翩翩起舞。

Luoqun (skirt made of thin silk)

It was made of a kind of silk gauze prevailed in Tang Dynasty (618-907). In this illustration, it depicts a lady in Tang Dynasty. She combs her hair in high buns wearing a round collar dress covered a red cappa. She wears *Luoqun* in two intervening colors on lower part of the body. She is dancing stretching her arms.

辽金元时期，汉族女性的裙装基本上沿袭了宋代的裙式，少数民族的裙装则保留了民族特点。如辽金时期的契丹、女真族穿裙，颜色多为黑紫色，上面绣全枝花，通常把裙穿在团衫内。

明代的裙式仍然具有唐宋时期裙装的特色。明初，女性以颜色素雅的裙子为尚，纹饰不明显。到了明末，裙上的纹样日益讲究，褶裥越来越密，出现了月华裙、凤尾裙等。传统的裙式在此时被进行了改制，如鱼鳞百褶裙就是在百褶裙的基础上发展并流行起来。

清代有一种朝裙，专供太后、皇后及命妇在祭祀、朝贺等典礼上穿用。此外，平民女子所穿的马面裙也是清代女裙中较有特色的样式。

some characteristics of Tang and Song Dynasties. Women in early Ming Dynasty wore plain skirt with little patterns on it. Until late Ming Dynasty, the number of patterns and folds increased greatly. Lunar corona skirt and phoenix tail skirt appeared. The traditional dress style was updated. Based on the pleated skirt, fish-scale pleated skirt was developed and flourished.

In Qing Dynasty (1644-1911), there was a court skirt exclusive for the use of the queen mother, empress and women who were given titles or ranks by the emperor when they offered sacrifice to gods or ancestors. In addition, horse-faced skirt worn by ordinary women had distinctive feature in Qing Dynasty.

- **凤尾裙**

  凤尾裙是一种由布条组成的女裙，将各色绸缎裁剪成宽窄条状，其中两条宽，余下的均为窄条，每条绣上花纹，两边镶滚金线，或者是缝缀花边。后部用彩条固定，上部与裙腰相连。因造型似凤尾而得名。穿着此裙时需要搭配衬裙，多为富家女子穿用。

  Phoenix tail skirt

  It's formed by strips of cloth. Colorful satins are cut into pieces among which two are wide others are narrow. Each strip has embroidery and gold thread edge piping on each end. Fixed by the ribbons on the back, the upper part is connected with waist of the skirt just like phoenix tail, hence its name. Underskirt must be worn to match it. It's mostly worn by women from wealthy family.

- 笼裙

笼裙是一种桶形的裙子，用轻薄纱罗为料制作，呈桶状，穿着时从头套入。最初常见于西南少数民族地区，隋唐时期传入中原。

Barrel-shaped dress

Barrel-shaped dress is made of light and thin gauze. The dress shapes a barrel pulling over the head. It appeared in southwestern ethnic minorities then it was introduced to the Central Plains during Sui and Tang Dynasties (589-907).

- 百褶裙

百褶裙是指有数道褶裥的女裙，褶裥布满周身，少则数十，多则逾百，常以数幅布帛为料制作。每道褶裥宽窄相等，并于裙腰处固定。隋唐时期，此裙多用于舞伎乐女，宋代时广为流行。

Pleat-dress

It refers to women's dress with many pleats on it. The number of the pleats ranges from several to more than a hundred. The dress is made of cotton and silk with the same width of pleat and fastened on the waist. It was mostly used among dancers in the court of Sui and Tang Dynasties (589-907). It became popular in Song Dynasty (960-1279).

- 马面裙

马面裙是清代最为常见且流行的裙式。裙的两侧是褶裥，中间有一段光面，俗称"马面"。常见的款式是在马面上缀以刺绣装饰，位置在马面中央或下端，四周并有镶边。

Horse-faced skirt

It's the most common and popular dress in Qing Dynasty. There are pleats on both flanks with a glossy satin in the middle, known as "horse face". The common style is to decorate with embroidery on central or lower part of the "horse face". It's surrounded by edge piping.

"拜倒在石榴裙下"

"拜倒在石榴裙下"是中国人熟知的俗语，多比喻男子对女子的爱慕倾倒之意。石榴与中国的服饰文化有着密切的联系，古代女子的裙装多为石榴红色，而当时染红裙的颜料也主要是从石榴花中提取而来。石榴裙在唐代非常流行，很多年轻的女子都喜欢穿着，显得格外美丽动人。石榴裙一直流传至明清，久而久之就成了古代年轻女子的代称。

相传唐明皇的宠妃杨玉环（又称"杨贵妃"）非常喜爱穿石榴裙。唐明皇过分宠爱杨贵妃导致终日不理朝政，大臣们不敢指责皇上，便迁怒于杨贵妃，对她拒不行礼。唐明皇感到宠妃受了委屈，于是下令：所有文官武将，见了贵妃一律行礼，拒不跪拜者，以欺君之罪严惩。大臣们无奈，只得在每次见到杨贵妃穿着石榴裙走来时，纷纷下跪行礼。于是，"拜倒在石榴裙下"的典故流传至今，演变成为男子爱慕、崇拜女子的俗语。

● 杨贵妃像
Portrait of concubine Yang

"Throwing Himself at Her Feet"

Literally translated as "bowed to a lady's pomegranate skirt" is a Chinese well-known saying. It's often a metaphor for a man's love to a woman. Pomegranate had close relations with Chinese costume. Women of ancient times mostly wore skirts in pomegranate-red because the dye was extracted from the pomegranate at that time. It's a very popular dress in Tang Dynasty (618-907). It was so beautiful and charming that many young women loved wearing it. It remained popular until Ming and Qing Dynasties (1368-1911) so it became the synonym of young lady in ancient times.

According to legend, Emperor Ming of Tang Dynasty beloved concubine Yang Yuhuan (also known as "concubine Yang") liked pomegranate skirt best. Emperor Ming of Tang Dynasty was so indulged in his affection to concubine Yang that he had no time to handle the state affairs. Because the ministers were intimidated by the emperor, they turned their resentfulness to concubine

Yang and refused to bow down to her. Emperor felt the wrongs she suffered and released an order: as long as meeting concubine Yang, the one who didn't bow down to her shall be given a severe punishment. Thus, the literary quotation "throwing himself at her feet" is passed down to present. It becomes the idiom to express the meaning of a man's affection to a woman.

● **石榴裙**

石榴裙是唐代非常流行的裙装，深受唐代女子喜爱。石榴裙的染料主要从石榴花、茜草中提取出来，颜色非常艳丽。画中的唐代仕女梳高髻，双肩披有一条印花帔帛，身穿高腰团花纹红裙，外罩宽大洒脱、轻薄透明的罗衫，肌肤隐约可见。

Pomegranate skirt

It was a popular dress which was loved by women in Tang Dynasty. The bright color is extracted from madder and pomegranate flowers. This painting depicts a lady in Tang Dynasty. She wears her hair in high buns and she wears a textile printing cappa, a high-waist floral red skirt, covering a loose outer dress. Her beautiful skin looms under the light and transparent silk dress.

绫罗绸缎

"绫罗绸缎"是一句成语，泛指各种精美的丝织品。"绫"是一种很薄的丝织品，一面光滑亮泽；"罗"是一种花纹紧密，纹理稀疏透亮的丝织品，可以形成花实地虚、明暗对照的视觉效果；"绸"是丝织品中最重要的一类，色泽鲜艳，手感平滑；"缎"是技术最为复杂、外观最为绚丽多彩的丝织品，质地柔软，平滑光亮，色彩丰富。

Ling Luo Chou Duan (Silk and Satin)

"Silk and satin" used as an idiom refers to a variety of fine silk fabrics. Chinese character *"Ling"* refers to a very thin silk with one glossy and smooth side. *Luo* is a kind of silk with translucent texture and close-knit patterns. It presents a strong contrast between light and dark. *"Chou"* is the most important silk fabrics. It's smooth and bright-colored. *Duan* feels smooth and soft. It is showy and colorful. This material is made by the most complicated technology.

- 丝织品
  Silk fabrics

首饰
Jewelry

中国古代的女子除了在服装上精心装扮，还要佩戴各种各样的首饰。高梳云髻，翠簪横插，耳垂珠串，丰富多姿的饰物成为古代女子扮美的点睛之笔。

In addition to decorating their clothes elaborately, ancient Chinese women wore various jewels. Ancient women wore their hair in high buns, hanging jewels on lobes. They wore various kinds of accessories which added colors for them.

古人除了创制出千姿百态的发式，还在美丽的发式上插戴各种饰品。笄、簪、钗、步摇等头饰，既能固定发髻，又能提升美感，是女子们必备的扮美饰品。

Ancient people created varied hairstyles and put various accessories on hair buns, such hairpins as *Ji* (large-sized hairpin), *Zan* (hairpin), *Chai* (hairpin), *Bu Yao* (hairpin with hanging beads) to fix the buns. They were essential ornaments for women.

笄不仅有固定和装饰发髻的作用,还是区分贵贱等级的工具。不同身份的人,所用的笄的材质也有所不同,帝后、诸侯的妻子用玉笄,士大夫的妻子用象牙笄,一般平民女子只能用骨笄。

Ji was used not only for ornament but also for distinguishing people's ranks. People in different ranks adopted different materials. Emperor, dukes, princes wore jade-made *Ji* and scholar-bureaucrat's wife wore ivory-made *Ji*. Bone-made *Ji* was used among common women.

汉代流行玉簪。簪还有一种特殊的形制,即簪花,一般用罗绢、通草或彩纸做成假花,精巧逼真,甚至达到乱真的程度。

Jade *Zan* was popular in Han Dynasty. It had another special type, called flower hairpin. The fake flowers were made of tough silk, grass and colored paper. It was made so delicate and lifelike that it was hard to distinguish between the fake and the real one.

发钗与簪的作用一样,都用来插发,只是簪作成一股,发钗则作成双股,形状像树枝枝桠。早在先秦时代,女子就已经佩戴发钗,此后一直流行。

Both *Zan* and *Chai* were used to insert in hair. *Zan* had one strand while *Chai* had two strands like a prong. Starting from pre-Qin era, women wore *Chai* which had been popular since then.

- 春秋时期的玉笄
  Jade *Ji* of the Spring and Autumn Period (770B.C.-476B.C.)

- 明代的镶宝石金簪
  Gold *Zan* (hairpin)inlaid with gemstone of Ming Dynasty (1368-1644)

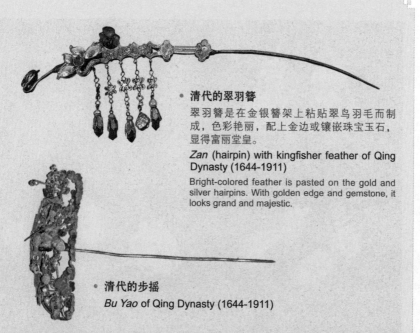

- 清代的翠羽簪

  翠羽簪是在金银簪架上粘贴翠鸟羽毛而制成，色彩艳丽，配上金边或镶嵌珠宝玉石，显得富丽堂皇。

  Zan (hairpin) with kingfisher feather of Qing Dynasty (1644-1911)

  Bright-colored feather is pasted on the gold and silver hairpins. With golden edge and gemstone, it looks grand and majestic.

- 清代的步摇

  Bu Yao of Qing Dynasty (1644-1911)

步摇也是古代女子的一种头饰，因钗首缀着活动的坠饰，会随着步履不停地摇曳而得名。汉代就很流行簪步摇，魏晋南北朝时尤为盛行。唐代以后，步摇多为金玉制凤鸟口衔串珠，称为"凤头钗"。明清时期也流行一种步摇，多在簪钗首垂珠坠饰，称为"步摇簪""步摇钗"。

Bu Yao was another kind of hair accessories. Because there was hanging jewelry on it, it was swaying with women's walking pace, hence its name. It was prevailed in Han Dynasty lasting Wei, Jin, Southern and Northern Dynasties. After Tang Dynasty (618-907), Bu Yao was made of gold and jade with the design of a bead in phoenix beak, known as "crested hairpin". Bu Yao with hanging jewels on hairpins prevailed in Ming Dynasty (1368-1644), known as "Bu Yao hairpin".

项链、耳环、手镯、戒指这些首饰，也是古代女子喜爱的装饰品，常被用作服装的配套装饰。

Such jewelry as necklaces, earrings, bracelets and rings was beloved by ancient women. They were used as ornaments to match costumes.

- 清代的银耳坠

  Silver eardrops of Qing Dynasty (1644-1911)

早在战国时期，就已经有了穿耳戴环的风俗。耳环普遍使用，始于宋代，复杂和贵重的耳环大都用金银玉石打磨制成，轻巧精美。明代以后，妇女穿耳戴环已是十分平常的事了。耳环的材料和造型更加丰富，人们用金银玉石、珊瑚玛瑙、珍珠贝壳、竹木等材料制成各种耳环，耳环上面的镶嵌之物极华贵而繁复，有宝石、水晶、玛瑙、珍珠等。耳环的称呼因材质、形状和构造的不同而多种多样：形状大的叫"耳环"；形状小的叫"耳塞"；耳环下面缀以垂饰的称"耳坠"。

As early as the Warring States Period (475B.C.-221B.C.), it's been the custom to pierce the ear holes for wearing rings. Earrings were widely used from Song Dynasty (960-1279). Gold, silver and jade were ground and polished to make complex and precious earrings. After Ming Dynasty (1368-1644), it's a common practice for women to pierce ear holes for rings. Materials and shapes of the earrings became more diverse. They included gold, silver, jade, pearl, coral, carnelian, shell, bamboo and so on. People extravagantly inlaid earrings with jewels such as gemstone, carnelian, pearl and crystal. The name of earrings varied from their shapes, earrings in big size named as "earring", in small size called "ear insert" and earrings with hanging pendants known as "eardrop".

手镯和项链起源于母系社会向父系社会过渡时期一些地区的抢婚制。当时，部落里的男子会定期组织一批人，将另一部落的姑娘强抢回来成婚。在抢的过程中，为了防止姑娘逃跑，就用绳子或链子捆住姑娘的脖子和手。随着社会的进步，抢婚习俗逐渐被淘汰，那些为了防止姑娘逃走的"链子"和"手铐"，变成了用黄金、白银、美玉来制作的装饰品，成为戴在手腕上的手镯，挂在颈上的项链或项圈。女子戴手镯和项链在唐代已经很流行，到了明清就非常普遍了。

• 清代的银手镯
Silver bracelets of Qing Dynasty (1644-1911)

Bracelets and necklaces were originated from the transitional period from matriarchal society to patriarchal society when they had the custom of marriage by capture. At that time, men regularly organized a group of people to seize women from another tribe for being married. In case of seized women escaping, they were shackled with ropes and chains. With the development of society, the custom of marriage by capture disappeared. Those chains and shackles were replaced by gold, silver, and jade accessories. Women wearing bracelets and necklaces prevailed in Tang Dynasty (618-907). It's become very common by Ming and Qing Dynasties (1368-1911).

戒指，又称"指环"，源于两千多年前的宫廷生活。中国古代女子戴戒指是很普遍的，用以记事，是一种"禁戒""戒止"的标志。古代皇帝有三宫、六院、七十二妃嫔，当后妃有月事或妊娠，宦官就给其一枚金戒指戴在左手上，以示戒身，皇上就不能与她亲近了。如果皇上看上了哪位宫女，宦官就记下她陪伴君王的日期，并在她的右手上戴一枚银戒指作为记号。后来，戒指逐渐变成女性的装饰品，并作为婚姻的信物。

• 清代的项圈
Chaplet of Qing Dynasty (1644-1911)

Ring was originated from imperial palace more than two thousand years ago. It's very common for imperial women to wear rings, which was used for keeping a note. It's also a symbol of "forbidden". Emperor of ancient times boasted three Inner Courts, six Palaces and seventy-two concubines. When concubines had pregnancy or in the period of that month, eunuch would give them a gold ring wearing on a finger of left hand to show "forbidden". Emperor wasn't intimate with them when seeing rings. If the emperor fell in love with a maid in an imperial palace, the eunuch would note down the date when she was with the emperor and she was given a silver ring wearing her finger of the right hand. Later it evolved into an ornament and acted as a marriage token.

• 清代金镶玉戒指
Gold ring inlaid with jade of Qing Dynasty (1644-1911)

> 裤

在中国服饰史上，传统服装一般为"上衣下裳"，或者上下相连，裤子大多以内衣的形式存在。在古代，"裤"被称为"绔"或"袴"。

春秋战国时期，人们就已开始穿裤，但其形制还不完备，被称为"胫衣"，胫衣不分男女均可穿用。穿胫衣的目的是为了保护小腿，但膝盖以上无遮护，外面还要穿裳遮盖。穿胫衣的缺点是行动不便，尤其不利于战争骑射。

战国时期，传统的胫衣被改为裤裆相连的合裆裤，用以遮蔽大腿，最初仅用于军服。到汉代时，合裆裤已流传至民间，为百姓穿用。为区别开裆的"袴"，合裆的裤称为"裈"。魏晋南北朝以后，"袴""裈"二字合用。裤裆被缝

> Pants

In the history of Chinese costume, the style of traditional dress was generally "upper *Yi* and lower *Shang*" or the upper part connected with the lower part of the clothes. Pants served as underwear. In ancient times, "pants" was referred to as *Ku*.

People started to wear pants from the Spring and Autumn and Warring States Period (770B.C.-221B.C.). However, the shape was not complete, known as "*Jing Yi*" at that time. Men and women wore it to protect crus but knees were exposed so *Shang* had to be worn over them. Wearing *Jing Yi* was not conductive to riding and shooting.

In Warring State Period (475B.C.-221B.C.), crotch was added on the traditional *Jing Yi* to cover thighs. It was initially used as military uniform. It spread among common people in

- 开裆裤

  开裆裤是一种裆部不缝合的裤子，无论男女，皆可穿着，形制是在胫衣的基础之上发展而成。裤管上连缀一裆，且裆部缝合，上连于腰，将有裆的一面穿在身后。明清时期仍有这种服饰，清中至后期，一般多用于儿童。

  Opening-crotch pants

  It's a kind of pants without sewn crotch. Men and women can wear it. It evolves from *Jing Yi*. Pant legs connect with crotch and waist then the sewn crotch is worn at back of the body. It remained to be used in Ming and Qing Dynasties. In mid and late Qing Dynasty, it was mainly for children.

- 胫衣

  胫衣是中国早期的裤子形式，与后世的套裤相似，是一种无腰无裆的裤管，穿时套在膝盖下沿，用绳带结系。

  Jingyi

  It's the inchoate shape of Chinese pants. *Jing Yi* is similar to the later leggings without waist and crotch. It's covered below the knees and tied with rope and ribbon.

合后，裤就可单独穿用，不必外加裳了。魏晋南北朝时期是裤装发展的高峰时期，受少数民族服装影响，还出现了袴褶。最初，袴褶多用作军服，后因其简便、合体，逐渐成为平民百姓的常服，无论男女都普遍穿着。袴褶有大口、小口两种形制，一般以布帛制成，为了御寒，也可絮进棉、麻等，称为"复袴"。

Han Dynasty (206B.C.-220A.D.). To distinguish the style of opening-crotch from co-crotch, they were written in different Chinese characters. Two characters, *Ku* and *Kun*, were combined together after Wei, Jin, Southern and Northern Dynasties (220-581). After opening-crotch was sewn, pants could be used independently. It was no need to wear additional over it. The development of pants reached the peak in Wei, Jin,

唐代的男子以袍衫为常服，袍衫之内穿裤，裤作为内衣，款式变化不大。唐代的女子喜欢穿裙，但裤没有被弃用，在"胡服"盛行之时，穿裤成为一种时尚。此时的裤管做得比较紧窄，裤脚也明显收束。

到了宋代，封建伦理观念兴盛，女子穿裤不能露在外面，而要用长裙掩盖，在胫衣的基础上形成膝裤、无裆裤。明清时期，膝裤仍然流行。明代的膝裤多制成平口，上达膝部，下及脚踝。清代称膝裤为"套裤"，长度不限于膝下，也有遮住大腿的。除了膝裤之外，长

Southern and Northern Dynasties (220-581). Influenced by the ethnic costumes, *Ku Zhe* appeared. It was mainly used as military uniform. Because of its simple design, *Ku Zhe* became informal dress for common people, which was for men and women. There were two types of *Ku Zhe* with either loose legs or narrow legs. It was made of cloth and silk. Cotton and flax were padded to keep warm, known as *Fu Kun*.

In Tang Dynasty, men took gown as their informal dress. Pants as underwear with little change in style, were worn inside gown. Women in Tang Dynasty preferred wearing skirt, but pants were not abandoned. During "*Hu* Dress" prevailing period, wearing pants became fashionable. At that time, pant legs and bottom of legs tended to be narrow.

- **穿袴褶的书童**

魏晋南北朝时期的裤有大口裤和小口裤之分。与大口裤相搭配的上衣较为紧身合体，称为"褶"。褶、袴穿在一起，称为"袴褶"。外面不用再穿着裳、裙。

A boy who serves in a scholar's study wears *Ku Zhe*

Pants in Southern and Northern Dynasties could be divided into large leg and small leg. A kind of tight coat matched large-legged pants, known as Zhe. Together with pants, they referred to as *Ku Zhe*. *Shang* doesn't have to be worn outside.

裤、短裤的使用都很普遍，裤身多宽大，如牛头、叉裤、灯笼裤等。男子把裤衬在袍衫之内，或与襦袄相搭配；女子则把裤穿在裙内。

The feudal ethics flourished in Song Dynasty. Pants worn by women were not allowed to expose, which shall be covered by a long dress. Based on the shape of *Jing Yi*, knee pants and non-crotch pants came into being. Knee pants remained popular in Ming and Qing Dynasties (1368-1911). In Ming Dynasty (1368-1644), knee pants were level-legged with the length from knees to angles. In Qing Dynasty (1644-1911), knee pants were also known as "Leggings" with the length beyond knees and the length extending to thighs. Besides knee pants, long pants and short pants became popular. Pants were large and loose, such as, ox head-shaped pants and bloomers. Men wore gown over the pants. They also matched pants with *Ru* and *Ao*. Women wore the skirt over the pants.

- **穿小口袴褶的男子**
  小口袴褶是袴褶的一种，衣袖及袴脚都制作得较为窄小，与大口袴褶相区别，南北朝时较为流行，多用于北方的少数民族。
  A man wearing small-legged *Ku Zhe*
  It's a kind of *Ku Zhe* with narrow and small legs which is different from loose style. It was popular among the northern ethnic minorities during Southern and Northern Dynasties (420-581).

- 膝裤

膝裤又称"套裤",两裤筒各自分开,不连成一体,没有腰,没有裆,上面系带,可束在胫上,上长至膝,下长至踝,用带子系扎。男女皆可穿,穿时加罩在长裤之外。裤管的造型多样,清初,上下垂直,为直筒状;清中期,上宽下窄,裤管下部紧裹胫上,裤脚开衩;清晚期,裤管宽松肥大。女子所穿的套裤,裤脚常镶有花边,色彩鲜艳。

Knee pants

Knee pants are known as "leggings" with two individual trouser legs. Knee pants are not a whole piece without waist and crotch. Shins are bound by ribbons. Leggings are tied with a ribbon around the waist. Both men and women can wear. Leggings are worn over long pants. There were many types of trouser legs; in early Qing Dynasty, there was upright in barrel shape; in mid Qing Dynasty, there was upper in wide and lower in narrow tied around the shins tightly; in late Qing Dynasty, there was a loose style. Leggings worn by women are mostly in bright-colored with laces on the bottom of legs.

"纨绔子弟"的由来

在中国古代,"裤"被称为"绔"或"袴"。贵族所穿的裤用丝绸等上等衣料制成,而平民百姓所穿的裤常用质地较次的布制成。明代张岱《夜航船·衣裳》载:"纨绔,贵家子弟之服。"成语"纨绔子弟"就由此衍生而来,用以形容衣着华丽、出身富贵,但不学无术的人。

Origin of "Fellows with White Silken Breeches – Fops"

In ancient China, "pants" was called *Ku*. Pants worn by nobility were made of fine fabrics while pants worn by common people were made of inferior cloth. In his literary works "Night Sailing, *Yi Shang*", litterateur Zhang Dai of Ming Dynasty (1368-1644) wrote "*Wan Ku* (white silken breeches) referred to the dress of young man from wealthy family." The idiom derives from it which means a well-dressed but good-for-nothing young man from a wealthy family.

> 背心

　　背心，又称"坎肩"或"马甲"，主要的特点是没有袖子或者袖子很短。魏晋时期的"裆"是背心的最早形制。在宋代时男女皆可穿背心，最初仅作为衬衣，后来可作为外衣，加罩在其他服装之外。夏季，也有一些男子只穿背心；冬季在背心内加棉絮，用于御寒。当时背心的制式多为对襟、直领，衣

> Sleeveless Jacket

Sleeveless jacket, also known as vest, features its short sleeves or non-sleeve. *Dang* in Wei and Jin Dynasties was the earliest shape of sleeveless jacket. *Dang* as underlinen was worn by women and men in Song Dynasty (960-1279), later it as outer garment was worn over other clothes. In summer, some men only wore sleeveless jacket. In winter, it was padded with cotton to keep warm. Sleeveless jacket at that time had opposite front

唐代的半臂

在古代服饰中，除采用长袖衣外，也有用短袖衣的。短袖衣的衣袖为长袖衣之半，故称为"半袖"，是一种很特殊的背心。唐宋时期，半袖多被称为"半臂"。半袖采用对襟，袖长至臂中，袖口有两种形式：一种宽大平直，没有装饰；另一种则加以缘饰，并施以褶皱。

Half-sleeved garment in Tang Dynasty

In addition to long-sleeved garments, there is short-sleeved garment in Chinese ancient costumes. Sleeves of this garment are half length of that of long sleeve garment, hence its name, which is a special type of vest. It's designed as opposite front pieces. There are two types of cuffs; one is large and straight without decoration, the other is pleated with edging.

长至腰部，两侧开衩。衣襟间不用纽带，穿时任其敞开。到了明清时期，背心的形制有所变化，穿时衣襟以纽扣相连。

pieces without ribbon around the waist. It had a straight collar. There was a slit on each flank of the vest. Reaching the waist, it was worn open at random. In Qing Dynasty (1644-1911), there was a change in style, which became a vest with buttons down the front piece.

- **穿裲裆的武士**

  裲裆是最早的背心，形制一般为前后两片，质地为布帛，肩部用皮质搭袢连缀，腰间用皮革系扎。裲裆服饰南北朝时十分流行，一直至唐宋以后。当时的裲裆男女均可穿用，有的用彩绣，有的纳有丝棉，是后世棉背心的最早形式。

  A warrior in *Liang Dang*

  Liang Dang is the earliest style of vest. It is made of cotton and silk with one piece in front and the other piece on the back. Leather agraffes are sewn on the shoulders. A leather tie binds around the waist. *Liang Dang* was popular from Southern and Northern Dynasties (420-581) to Tang and Song Dynasties (618-1279). There's colored embroidery on *Liang Dang*, which is worn by women and men. Some of them are padded with cotton, which constitutes the earliest form of cotton-padded vest.

- **比甲**

  比甲的外形类似背心，是流行于元、明、清时期的便服。其特点是对襟，直领，下长过膝，穿着时罩在衫袄之外。因为穿着方便，受士、庶阶层的女子欢迎。

  Bi Jia

  It looks like a vest from its appearance. It was an informal dress popularized in Yuan, Ming and Qing Dynasties (1271-1911). It features opposite front pieces, straight collar and the length reaching beyond knees. It's worn over *Ao*. *Bi Jia* is popular among common women and scholars for its convenience.

- **巴图鲁坎肩**

  巴图鲁坎肩流行于晚清，意出满语"巴图鲁"（即勇士）。巴图鲁坎肩是一种多纽扣的背心，男女皆可穿用，通常制成胸背两片，无袖，长不过腰，在前胸处横开一襟，上钉纽扣七粒，左右两腋各钉纽扣三粒，合计十三粒，因而也称"一字襟马甲""十三太保"。

  ### *Batulu* style vest

  Batulu is a Mongolian word meaning warrior. *Batulu* style vest prevailed in Qing Dynasty. It's a multi-button vest which can be worn by men and women. There is one piece on the chest and another piece on the back. Its length reaches above the waist. There's a straight line open on the chest where stand seven buttons. Another three buttons are sewn on each side of armpit. There are thirteen buttons in total, also known as "vest with buttons on a straight line" or "thirteen buttons vest".

> 内衣

中国服饰史上，不同时期的内衣有不同的称谓，汉代称为"心衣"，唐代出现了一种无带的内衣，被称为"诃子"，宋代出现了"抹胸"，元代出现了"合欢襟"，明代出现了"主腰"，清代的女性内衣则称"肚兜"。

唐代以前的内衣都是有肩带的，而到了唐代，女性以丰腴为美，喜爱将裙子高束在胸际，于是发明了无带内衣，即"诃子"。诃子仅在胸部系一根宽带子，两肩、上胸和后背袒露，外披透明的罗衣，内衣若隐若现。

清代非常流行的肚兜也是一种极有特色的内衣。肚兜一般做成菱形，也有长方形、正方形、如意形、扇形、三角形等。菱形肚兜采

> Underwear

In history of Chinese costumes, the name of underwear varied in different period of time. In Han Dynasty (206 B.C.-220 A.D.), it was referred to as "heart clothes". A kind of underwear without tie appeared in Tang Dynasty (618-907) which was named as *He Zi. Mo Xiong* appeared in Song Dynasty (960-1279). *He Huan Jin* showed up in Yuan Dynasty (1271-1368). In Ming Dynasty (1368-1644), *Zhu Yao* emerged. Women in Qing Dynasty (1644-1911) named underwear as *Du Dou*.

There were aiguillettes on the underwear ahead of Tang Dynasty. In Tang Dynasty, beauty was described as a woman with full and round figure. Women preferred tying the dress around their breasts. The underwear without aiguillette was invented accordingly, that is, *He Zi*. A broad strap was tied around

• 宋代穿抹胸的女子
Lady in Song Dynasty wearing *Mo Xiong*

用对角设计，上角裁去，呈凹状浅半圆形，上端有带，穿时系在脖子上，左右两角各有一条带子，束在背后。下角呈圆弧形、三角形或尖形，遮过肚脐，直至小腹。肚兜

the breast exposing the shoulders and upper breast. *Luo Yi* (a kind of garment made of gauze) was worn over it to make underwear loom.

Du Dou (a kind of underwear loose covering for the breast), a distinctive underwear, was popular in Qing Dynasty (1644-1911). It's mostly in diamond shape such other shapes as rectangular, square, triangular, *Ruyi* shape, fan shape and so on. Diamond- shaped *Du Dou* was designed as opposite angles. Cut off

遮盖住了胸部和肚脐，起到遮羞和保暖的功能。同时，独特的菱形结构，增添了视觉美。

the upper angle, it appeared a shallow semicircular with concavity. There were ribbons on its top and it was tied around the neck. There was another ribbon on each side of the other two angles which was tied around the back. Lower *Du Dou* presented in circular shape, triangular shape and closed angle. Because it veiled the navel, belly and breast, it had the function of health care and covering up women's embarrassment part of the body. Meanwhile, the design of diamond-shaped structure aroused pleasant sensation.

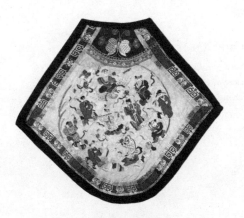

- 肚兜

清代的女子非常重视肚兜，直到新婚之夜才能被丈夫看到。除自己外，女子一生只为情人和孩子绣肚兜。

Du Dou

Women in Qing Dynasty attached great importance to *Du Dou*. It wouldn't be seen by others except for her husband until the first night of the wedding. Women could only embroider *Du Dou* for her lover and her son besides for herself.

内衣与"亵衣"

古人对待内衣，往往抱着回避和隐讳的态度，因此内衣最早被称为"亵衣"。"亵"即意为轻浮、不庄重，所以内衣是不可以轻易示人的，否则便有失体统。如果女人在人前露出内衣，那一定不是良家女子。

Underwear and *Xie Yi*

Ancient people mostly avoided mentioning underwear because they took it as *Xie Yi*. Character *Xie* has the meaning of flightiness. It's not a proper behavior for ancient people to have their underwear exposed to others. If a lady exposed her underwear in front of others, she was deemed to be in bad virtue.

> 巾、帽

　　巾和帽都是中国古代的首服，戴在头上，与衣裳相配。最早的巾和帽以保暖御寒为主要目的，多由质地厚实的布帛或毛皮制成。随着社会制度和人文礼教的发展，冠服制度被纳入"礼制"的范畴，成为区分等级、辨别尊卑的重要标志。上自帝王后妃，下至平民百姓，首服的形式与材质各不相同。一般来说，帝王后妃戴冠，而平民百姓则戴巾或帽。

　　巾是裹头用的布帕，最初的作用只是擦汗，后来则用来裹头，一物两用。帽是在巾的基础上演变而成的，帽与巾的区别在于缝合与否。由于戴帽比扎巾方便省事，便逐渐取而代之。帽的最初用途主要是为了御寒，多以毛毡制成。

> Scarf and Hat

Scarf and hat were the ornaments for ancient Chinese. People wore them on head to match the dress. The earliest scarf and hat were made of heavy cloth or leather to keep warm. With the development of social system and human ethical code, Crown and Costumes became part of "social etiquette". It's an important symbol to distinguish people's ranks and social status. From the emperor and his concubines down to ordinary people, their head ornaments varied in shape and material. In general, the emperor and imperial concubines wore crown while the civilians wore scarf and hat.

　　Scarves were made of cloth originally used as wiping sweat, later used as wrapping around the head. Hats evolved from scarves. The difference between hats and scarves lied in whether

• 明末的文人雅士像
Portrait of literatus in the late Ming Dynasty (1368-1644)

儒巾：明代儒士、士人常戴的一种头巾，以漆藤丝或麻布为里，黑绉纱为表。帽围呈圆形，巾身由四片布帛缝制而成，顶部四角隆起，呈方形，后有两条垂带。

Ru Jin was a scarf worn among scholars in Ming Dynasty. It was in round shape with vine and linen inside and crape covering the surface. The body of *Ru Jin* was sewn in four pieces of cloth. Four top corners rose in square shape dropping two bands on the back.

秦汉以前，平民百姓多以布裹头，称为"帻"，颜色以青、黑为主。此时的帽以西域少数民族所戴为多，而中原地区戴帽者，主要是孩童。

东汉末年，由于统治者的使用和士人的不拘于礼法，轻便的头巾大为流行，扎巾成为时髦装扮，王公贵族也用头巾来束发，以附庸风

it's sewn up or not. Wearing hat was much easier than wrapping scarf around the head so the scarf was gradually replaced. Hats were most made of hair felt to keep warm.

Ahead of Qin and Han Dynasties (206B.C.-220A.D.), common people wrapped their heads with black and cyan cloth. At that time, hats were worn by the ethnic minorities in the Western Regions.

四方平定巾：又称"方巾"，是明代流行的一种首服，为读书人和平民所戴。以黑色纱罗制成，可以折叠，展开时四角皆方，故名。

Si Fang Ping Ding Jin is also known as "Square Scarf". It was worn among literatus and common people prevailed in Ming Dynasty. It's made of black gauze which is either foldable or expandable in square shape.

雅，称为"幅巾"。幅巾通常用布帛裁成方形，长宽和布幅相等，使用时通常以幅巾包裹发髻，在额前或颅后系结。当时的巾为方帕，每次使用时都要临时系裹，后来则出现了事先被折叠缝制成各种形状的巾，用时直接戴在头上，非常方便。东汉时出现的角巾就属于这种形制。逐渐地，帽的式样也多了起

In the Central Plains, only children wore hats.

At the end of Eastern Han Dynasty (25-220), because scarves were widely used among the dominator and scholars irrespective of restriction, they became even more popular. Princes and aristocrats used scarves as hairdo known as *Fu Jin* to follow the chic deliberately. *Fu Jin* was made of a square piece of

来，越来越受人们喜爱。

魏晋南北朝时期，幅巾更为流行，除了家居时使用外，还可以用于礼见。北周武帝时期，对幅巾做了改进，将方帕裁出四角，用以裹头，其两条系于颅后垂下，两条反系在头上，这种形制的头巾称"幞头"。此时，帽子在中原地区得到了普及。此后，随着戴帽者的增多，帽子的作用已不限于御寒，式样不断更新，材质也多样起来。南北朝时期，一种以纱制成的纱帽是男子的主要首服。纱帽分两种颜色，白色多为贵族所戴，黑色多为百姓所戴。

隋唐时期，幞头是男子主要的首服。幞头比较软，不太美观，人们便在幞头里垫入衬物，使幞头显得更硬挺，这个衬物称"巾子"。到了五代，幞头发生了较大变化，通常做成方裹型，顶部有两层，前低后高，形似帽子。隋唐时期的帽沿用南北朝的式样，纱帽的使用仍很普遍。笠帽在当时也很普遍，以竹篾、棕皮、草葛及毡类等材料编成，形状多为圆形，有宽大的帽檐，用于遮阳、避雨，常用于劳作

cloth. Hair buns were wrapped by *Fu Jin*, tied on the forehead or on the back of the head. Scarf at that time was in square. People had to wrap and tie each time when they wore scarves. Later lapped sewn scarves in different shapes appeared which were easy to use. *Jiao Jin* of Eastern Han Dynasty (25-220) was one of them. As hats became more diverse, they were beloved by more and more people.

During Wei, Jin, Southern and Northern Dynasties (220-581), *Fu Jin* became more popular, used not only at home but also in ceremonies. During Emperor Wu of the Northern Zhou Reign (560-578) *Fu Jin* got improved. Four strips were tailored out of a square piece of cloth to wrap the head. Two strips were tied at back of the head dangling and the other two "feet" were tied on the opposite direction. This kind of scarf was known as *Fu Tou*. During that time, hats became popular in the Central Plains. With the increasing number of people wearing hats, hat styles were updated constantly and material varied. Thus, hats were not limited to keep warm. In Southern and Northern Dynasties, a kind of hats made of gauze was used as men's wear. This type of hat had two colors:

- **明代戴唐巾的男子**

唐巾，又称"软巾"，以乌纱制成的头巾。形制与唐代的幞头相似，区别在于其后下垂二角，里面纳有藤篾，向两旁分开，呈八字形。宋元时期较为流行，通常用于便服，男女尊卑均可戴用。

A man wearing *Tang Jin* in Ming Dynasty (1368-1644)

Tang Jin also known as "soft scarf" is made of black gauze. Its shape was similar to *Fu Tou* of Tang Dynasty (618-907) with difference in hanging two strips of cloth. There's a vine in *Tang Jin* to extend in spread-angled. It was used as informal dress in Song and Yuan Dynasties (960-1279), which could be worn by anyone regardless of social status.

white was worn by aristocrats, black was for the civilians.

During Sui and Tang Dynasties (581-907), *Fu Tou* was men's wear. After padded with infilling (known as *Jinzi*) to make it strong, *Fu Tou* didn't look as soft as it used to be. *Fu Tou* underwent great changes in the Five Dynasties (907-960). It looked like a square-shaped hat with two layers on the top tilting in front. In Sui and Tang Dynasties (581-907), hats followed the styles in Southern and Northern Dynasties (420-581). Gauze hats were still popular. *Li Mao*, woven by bamboo, grass and felt, was commonly used at that time. The round shape *Li Mao* was very common, which had a big and wide brim to keep away from sunshine and protect people from getting wet when they worked in the field. Custom of Han ethnic group was deeply influenced by Western ethnic minorities in Tang Dynasty (618-907). *Hu Mao* which was worn by the ethnic minorities was introduced into the Central Plains and prevailed at that time.

Fu Tou was used in Song Dynasty (960-1279) and evolved into the official hat for the emperor and officials. Scholars and persons with literary reputation advocated using *Fu Jin* to wrap around

● 五代时戴纱帽的男子
图中男子所戴的纱帽以黑色细纱制成，形状高耸。
Men in the Five Dynasties (907-960) wearing gauze hats
In this illustration shows men wearing tall and erect gauze hats.

之人。唐代汉族服饰受西域服饰的影响很大，少数民族所戴的胡帽传入中原，在当时颇为流行。

宋代仍沿用幞头，并发展成为帝王百官的官帽。此时的文人、名士又崇尚起以幅巾裹头的习惯。宋代幞头形制多样，幞头的顶有方圆之分，以方型为主流；角有软硬之分。由宋至元的四百年间，扎巾的习俗经久不衰。宋代士人戴帽风气

heads. *Fu Tou* was varied in style. Square-shaped *Fu Tou* was a mainstream but there was round-shaped *Fu Tou* with distinction of hard strips and soft strips. During four hundred odd years from Song to Yuan Dynasties (960-1368), *Zha Jin* prevailed. People in Song Dynasty (960-1279) not only liked wearing hat but also created lots of new styles. By Yuan Dynasty (1271-1368), hats worn in winter were mainly made of leather,

极盛，且别出心裁，样式繁多。到了元代，冬天戴的帽以皮质为主，称"暖帽"。暖帽形似风帽，帽檐上翻，帽顶为圆形或桃形，帽后披搭在肩上。夏天通常戴笠式帽，笠式帽因形制与笠帽相似而得名，是元代蒙古族特有的帽式。

明代裹巾的风气有增无减，超过了以往任何时代，其中以网巾和四方平定巾使用最广。明代男子所戴的帽种类繁多，普通男子大多戴一种名为"六合一统帽"的圆帽。清代男子日常的首服沿用六合一统帽，形制略有变化，有的为尖顶，有

known as "winter hat". The brim of winter hat was upturning. The top of the hat was in round and heart shape. The back of the hat hung on the shoulders. *Li Shi Mao* was worn in summer. *Li Shi Mao* and *Li Mao* were similar in shape, hence its name. It's the Mongolian unique style in Yuan Dynasty.

There's increasing trend to use *Guo Jin* in Ming Dynasty which exceeded any other dynasties. *Wang Jin* and *Si Fang Ping Ding Jin* were widely used at that time. Men in Ming Dynasty wore various types of hats. Ordinary men mostly wore *Liu He Yi Tong* Hat which followed in Qing Dynasty with a little variations in its

- **帷帽**

  帷帽是在藤编成的笠帽上再装一圈纱网，可起到屏蔽的作用。此帽本是西北少数民族女子遮阳、防风之用，后传入中原，成为汉族女子骑马出行的装扮。此帽在唐代武则天朝尤为盛行，因为其简洁轻便、戴卸方便，并且可以将脸浅露在外，深受当时女子喜爱。

  Wei Mao (curtained hat)

  Based on rattan mat woven *Li Mao*, *Wei Mao* was formed by adding a circle of gauze to *Li Mao* for the purpose of shielding. It's originally used among the ethnic women in the Northwest of China to protect again wind and sunshine. Later when it spreads to the Central Plains, it becomes women's horse riding custom. Because the hat slightly unveils the face and it's also convenient for wearing and taking off, it's very popular with women.

的为平顶，帽边有宽有窄。帽顶常装饰一颗结子，有的帽额上缀有玉石。这种帽分瓣明显，形似西瓜。

shape and structure. Some had acuminate top and others had flat top with a seed as decoration. Hat brim varied in narrow and broad. Stitched jades were found on the front of some hats. The shape looked like a watermelon.

- **戴六合一统帽的男子**

  六合一统帽，取六方统一之意。最早见于元代，明清两代沿用，是男子普遍使用的一种帽。制作时先将布片剪成六瓣，然后缝合在一起，形如倒扣的瓜皮，俗称"瓜皮帽"。此帽一般用作便帽，冬春以缎为材料，夏秋以纱为材料。

  A man with a *Liu He Yi Tong* hat (a kind of skullcap)

  "*Liu He Yi Tong*" means unification of the nation. This hat originally appeared in Yuan Dynasty (1271-1368) and evolved in Ming and Qing Dynasties (1368-1911) worn by men. Six pieces of cloth are sewn together presenting the shape of a inverted watermelon, commonly known as "watermelon cap". Winter hat is made of satin while summer hat is made of silk.

- **清代包髻的女子**

  包髻是一种长方形头巾，戴时将对角折叠，从额前向后面缠裹，再将巾角绕到额前打结。宋代就已出现此头巾。清代，由于剃发令的实施，男子已不扎巾，巾只在女子间流行。

  A woman in Qing Dynasty (1644-1911) with a scarf wrapped her hair buns

  It was a kind of long scarf folded in opposite angles. Starting from the forehead, it was wrapped around the hair buns and back to forehead to tie a knot. This type of scarf appeared in Song Dynasty (960-1279). By Qing Dynasty (1644-1911), with the implementation of cutting off men's plaits, men had no long wrapped scarf around hair buns. From then on, it was only popular among women.

面衣

面衣是用来覆面的巾，通常以绫罗为材料。在使用面衣时，除眼和口鼻部露出之外，其余均被蒙住，通常是出门远行的人所穿用，以抵御寒冷。

Mian Yi

Mian Yi was a kind of silk scarf to cover the face. People who made long distance travel usually wore it to cover the face with exposure of mouth and nose to keep warm.

- 唐代着面衣的贵族
  An aristocrat wearing *Mian Yi* in Tang Dynasty (618-907)

秋板貂鼠昭君套

昭君套是用条状貂皮制作的头饰，围在发髻之下、额头之上，是一种没有顶子、头部中空的帽套。昭君套相传为王昭君（中国古代四大美女之一，汉代人）出塞时所戴，故名，明清时期北方妇女在冬季多戴昭君套御寒。

秋板貂鼠是立秋至立冬前在貂鼠身上剥取的毛皮。貂鼠即紫貂，毛皮颜色有白色、纯黑和褐色三种，浓密柔滑，挡风性能强，不易浸湿，多制成衣服和围巾，有"裘中之王"的美称。由于貂鼠毛皮产量极少，价格昂贵，又被称为"软黄金"。

Autumn marten fur and *Zhao Jun Tao* (Mink Fur around Hair Buns)

Zhao Jun Tao made of a strip of mink fur was tied around hair buns. It was a kind of loops without top and the top of the head was exposed. According to legend, *Zhao Jun Tao* is designed for Wang Zhaojun coming out to border area. (She was the Han nationality, one of the four beauties in ancient China.) *Zhao Jun Tao* was named after her. Northern women wore it to keep warm in Ming and Qing Dynasties (1368-1911).

Marten fur was selected from autumn begins to the beginning of winter. Marten fur has white, black and brown colors. Considering this fur had the advantage of thickness, keeping away wind and water proof, thus enjoying the reputation of "King of the fur", it was used to make clothes and scarves. The price for rare marten fur was very expensive so it was known as "soft gold".

● 王昭君像
此画中，王昭君头戴华丽的昭君套，身穿皮毛大衣，坐在虎皮垫上。

Portrait of Wang Zhaojun
This portrait depicts Wang Zhaojun wears splendid *Zhao Jun Tao* and fur overcoat sitting on a tiger skin pad.

"巾帼不让须眉"

巾帼是中国古代贵族女子的头巾和发饰，宽大似冠、高耸显眼。中国常有"巾帼英雄""巾帼不让须眉"的说法，意指女子才华超群，胜过男子。这里的"巾帼"是褒义词，赞扬女子的英雄气概。

先秦时期，男子也戴巾帼。从汉代开始，巾帼才为女性专用，并成为代称。汉代在举行祭祀大典等重大活动时，宫廷贵妇都要戴一种用丝织品或发丝制成的头饰，罩在前额，勒于后脑，这便是"巾帼"，其上还装缀着一些金银珠玉制成的珍贵首饰。巾帼不易保存，后世早已弃用。长期以来，人们虽知此词，却未曾见过实物。

- 中国古代的巾帼英雄

此图塑造了四个不同时期的女英雄，执枪而立者为梁红玉（宋代抗金女英雄），执拂者为红拂女（隋末唐初时的侠女），背剑持盒者红线女（唐代武艺精湛的侠女），胸悬镜而旁立者红绡女（唐代勇敢追求自由与爱情的女子）。

China's heroines in ancient times

It depicts four heroines in different times. Liang Hongyu (anti-Jin Dynasty heroine in Song Dynasty (960-1279)) is holding a spear. Lady Hongfu (a chivalrous woman at the end of Sui (581-618) and the beginning of Tang Dynasties (618-907))is standing there holding Hongfu. Lady Hongxian (a chivalrous woman in Tang Dynasty (618-907))is holding a box with a sword on her back. Lady Hongxiao (a servant who pursues love and freedom courageously in Tang Dynasty (618-907))is standing there with a hanging mirror in front of her chest.

Women's superior talent is better than men's

Jin Guo referred to women's hair ornament and scarf in ancient China. It had the same width as crown and stood out. As a Chinese saying goes "*Jin Guo* heroine" or "*Jin Guo* is not inferior to any men." meaning women's superior talent is better than men's. *Jin Guo* is a commendatory term to speak highly of women's heroism.

Ahead of pre-Qin period, men wore *Jin Guo*. It was for women's exclusive use from Han Dynasty (206B.C.-220A.D.) thus *Jin Guo* became the synonym for women. When there's sacrificial ceremony held, the court ladies would wear a kind of ornament made of fabric or hair dotted with such precious jewels as gold, silver and jade to cover their foreheads and drag down to the back of the heads. It was named *Jin Guo*. It was not easy to save so it's no longer in use. Although people have been familiar with this word, no one has ever seen the physical *Jin Guo*.

"戴绿帽"的忌讳

"戴绿帽"是中国的一种俗语,用以讥刺妻子有外遇或淫行的丈夫。关于"戴绿帽"的起源,民间有很多种说法。普遍认可的说法是,古时候有一对夫妻,妻子长得娇艳可人,丈夫经常到外地去做生意。在丈夫外出的日子里,妻子不安寂寞跟街市上一个卖布的暗中交往。后来,妻子用绿色的布料做了一顶帽子给丈夫,并跟情人约定,每当看见丈夫戴上绿帽子外出,就可以相会了。

相传唐代的李封任延陵令时,如果下属官吏犯罪,不加杖罚,而是令其头裹绿头巾以作羞辱,期满后方能解下。至元明时,更是规定娼妓的亲属男子都要戴绿头巾。

Taboo of "Wearing Green Hat"

"Wear green hat" is a Chinese idiom to satirize a husband whose wife has affairs with others. Regarding the origin of this idiom, there are many variants among people. It's commonly believed this story. There's such a couple in ancient time. When her husband goes out for business, his beautiful wife who can't bear loneness will have affairs with a man who sells cloth at the market. His wife makes a green hat for his husband. As agreed with her lover, every time when he sees her husband going out with the green hat, they can date.

Legend has it that in Tang Dynasty (618-907) when Li Feng served as magistrate in Yan Ling he ordered any subordinate officials who committed crimes would not be punished by flogging instead of wrapping green scarf to be humiliated. It would be taken off after completion of his term of punishment. In Yuan and Ming Dynasties, prostitute's male family relatives shall wear green scarf.

> 鞋、袜

中国古代人对于穿鞋穿袜十分讲究，从鞋来说，主要有履、靴等。就质地而言有草编、绳织、丝织、木制、皮制等；就样式而言有方头、圆头、尖头等；就穿鞋的讲究而言，不同的历史时期有不同的礼节，不同的身份有不同的要求。足部的装饰作为服饰的一部分，同样反映着社会文化的发展历程。

古人日常生活中所穿的鞋子称为"履"，一般用丝帛制成或用丝帛装饰。商周时期，履多用葛麻或皮革制成，由于当时丝帛十分珍贵，所以纯粹以丝织品制成的鞋履尚不多见。

汉代，丝履的使用普遍起来，一些家境富庶的人家连奴婢都穿丝履。魏晋南北朝时，南方人多穿丝

> Shoes and Socks

Ancient Chinese people were very particular about shoes and socks. Shoes and boots were made of different textures, such as straw-woven, rope-woven, silk- woven, wood and leather etc. In terms of shape, it had round toe, square toe and pointed toe. Regulations on shoes wearing varied in different historical periods, people's different social status and different social institutions. Shoes decoration as part of costume and accessories reflected the social and cultural development.

Shoes worn by the ancient Chinese were known as *Lv* generally made of silk and cotton. Because silk and cotton was precious in Shang and Zhou Dynasties(1600B.C.-256B.C.), the shoes were mostly made of leather and flax instead.

Silk shoes became popular in Han

● 汉代的丝履

这双丝履全长26厘米，头宽7厘米，后跟深5厘米。以丝帛为面，为两层，鞋底用麻线编织。

Silk *Lv* (shoes) of Han Dynasty (206 B.C.-220 A.D.)

This pair of silk *Lv* was 26cm in length with toe 7cm and heel 5cm. The upper had two layers made of silk and cotton. Sole was woven by twine.

● 织成履

织成履用彩丝、棕麻等为料，依照事先定好的式样直接编成。考究者往往在鞋面上织有繁复的纹样，并有专业的工匠艺人织造织成履。此种履始于秦汉，魏、晋、南北朝时广为流行，多用于贵族男女。

Woven *Lv*

Based on fixed patterns, people wove the colored silk thread and linen into shoes. Someone is dainty about the complex patterns. They will be woven on the upper of the shoes and the professional craftsmen will make into shoes. This kind of shoes starts from Qin, Han, Wei and Jin Dynasties (221B.C.-420A.D.). In Southern and Northern Dynasties (420-581), it becomes popular among the noble.

Dynasty (206B.C.-220A.D.). Even the servants who served the wealthy family wore silk shoes. During Wei, Jin, Southern and Northern Dynasties (220-581), southern people mostly wore silk or stain shoes while the northern people wore them in summer. In winter, northern people changed the felt or cotton shoes. There were various shapes of silk shoes at that time. Shoes making technique became increasingly elaborate. There were embroidery patterns or different designs cut by gold foil stitched on the shoes. Shoes at that time had heavy sole which was different from it used to be.

In Tang Dynasty (618-907), the shape of silk shoes was followed and new structures constantly came out. The popular shoes featured their toes sticking up. At that time, the word *Lv* was replaced by "shoes" which has been in use today.

The court of Song and Yuan Dynasties (960-1368) set up the "silk shoes bureau" which was responsible for making shoes. Shoes making workshops and shoes stores emerged among the people. Women in Song Dynasty (960-1279) followed the fashion of little shoes and bound their feet to meet the small size of shoes which was known as

• 穿平头履的秦始皇陵兵马俑复原图

It's a restored illustration of terra-cotta warrior and horse at the tomb of Qin Shi Huang. This illustration shows terra-cotta warrior with flat toe shoes.

平头履：鞋头不高翘，有方形、方圆形、圆形三种样式。

Flat toe shoes: There are three styles-square, round and shape in round and square combined. The toe doesn't stick out.

履或缎履；北方人则在夏季穿丝履，冬季穿棉履或毡履。当时的丝履形制较为丰富，制作也日趋精细，如在上面织绣花纹，或是把金箔剪成各式图案，缝缀在丝履上。履底也有变化，以往的履（除舄之外）一般用的是薄底，而此时的鞋履采用厚底。

唐代沿袭前代的丝履形制，并不断推陈出新，比较典型的特点是翘头履的盛行。但此时，"鞋"字

"lotus feet".

In Ming and Qing Dynasties (1368-1911), men's shoes were made of shiny and heavy satin. Leather inlaid on the toe and heel part of the shoes, thus it was also called "inlaid shoes" which was not only pretty but also durable. Women's footwear was mainly made of colored satin. Young women's bright-colored shoes were embroidered all kinds of patterns. Compared with the previous dynasties, women's foot-binding shoes changed greatly with high-soled shoes

- 笏头履

笏头履是一种造型为高头的鞋履，履头高翘，呈笏板状，顶部为圆弧形。笏头履始于南朝，男女均可穿着，隋唐时期多用于女子，五代以后逐渐消失，明代时又有兴起，但形制有所变化，多用于男子。

Hu Tou shoes

Hu Tou shoes with raised toe have the similar shape as Huo plate, that is, a circular arc on the top. Hu Tuo shoes appeared in Southern Dynasty. Both men and women could wear. Women mostly wore it in Sui and Tang Dynasties (581-907). It gradually disappeared after Five Dynasties (907-960). It came back in Ming Dynasty (1368-1644) with variation in its shape, but mostly for men to wear.

- 穿歧头履的女子

歧头履是头部分歧的鞋履，鞋头做成两个尖角，中间凹陷，男女均可穿着。歧头履始于西汉，唐代仍有此制，宋代以后逐渐消失。

A woman with Qi Tou shoes

Toe cap Qi Tou shoes is made into two closed angles sunken in the middle. Both men and women can wear it. It started from the Western Han Dynasty (206B.C.-25A.D.).It's also found in Tang Dynasty (618-907) but it disappears after Song Dynasty (960-1279).

- 重台履

重台履是指高底、头部上翘的鞋履，始于南朝宋时，唐代女子也喜好穿着。

Zhong Tai shoes

It referred to high-soled and toe raised shoes. Starting from Southern Song Dynasty, women in Tang Dynasty also loved wearing it.

代替了"履"字，成为鞋子的通称，一直沿用至今。

宋元时期，官廷内专门设置了"丝鞋局"，负责管理、制作丝鞋。民间还出现了生产鞋履的作坊

as a main characteristic in Qing Dynasty (1644-1911). Because there was no custom for Manchu women to bind feet, their shoes tended to be large and loose compared with Han women's bow-shaped shoes. Manchu women's shoes were also

和销售鞋履的店铺。宋代女子缠足成为风气，因此女子所穿鞋履以纤细小巧为尚，俗称"三寸金莲"。

明清时期，男鞋多以质地厚实、富有光泽的缎子制成，在鞋头、鞋跟部位镶嵌皮革，称作"镶鞋"，既美观又结实耐用。女子的鞋履则多以彩缎制成。年轻女性的鞋履用色鲜艳，鞋上还刺绣有各式纹样。清代缠足女子穿的鞋履较前代发生了新的变化，最突出的特点是采用高底。因没有缠足的陋习，满族女子穿的旗鞋与汉族女子的弓鞋相比，显得特别宽大，也采用高底，因鞋踏在泥地上，印痕似马蹄，俗称"马蹄底"，鞋的造型似花盆，又称"花盆鞋"。高底的花盆鞋多为年轻女子穿用，随着年龄增长，中老年女性所穿的鞋，鞋底高度逐渐降低，以致改成平底。

除了鞋履，中国古代还流行穿靴，即高至踝骨以上的长筒鞋，多用皮制。古人穿皮制鞋子的历史由来已久，早在先秦时期，北方少数民族就用皮鞋御寒，汉族则多用于军旅，着靴者以兵士居多。从隋唐开始，靴子流行起来，被用作百官

high-soled. Considering the footprint on the mud presented the shape of horse's hoof, it's referred to as "horse's hoof sole" or vase-shaped shoes. High-soled vase-shaped shoes were mainly worn by young women. Height of sole was gradually decreased and tuned to be flattie for the elderly wear.

In addition to shoes, boots were also popular with ancient Chinese. Boots were mainly made of leather with the length above ankles. It was time-honored for ancient people to wear leather-made shoes. It was dated back pre-Qin period when northern ethnic minorities wore leather shoes to keep warm. In Han Dynasty (206B.C.-220A.D.), the majority of soldiers wore boots when they marched. Boots became popular from Sui and Tang Dynasties (581-907) which were used as officials' informal wear. Except on the certain occasions such as sacrificial offering ceremonies, celebrations and major events at the court, from the emperor down to the officials all wore boots in accordance with the current fashion. Boots of Tang Dynasty (618-907) were made of dyed black leather and sewn the varied sizes together. From Song to Ming Dynasties (960-1644), boots kept the design of

常服，除了祭祀、庆典、朝会等重大活动中仍使用舄外，一般都以穿靴为尚。上自帝王，下至百官，莫不如此。唐代的靴子，一般用染成黑色的皮料制成，即"乌皮靴"，由多块大小不等的皮料缝合而成。宋代至明代沿袭前制，靴子用于朝服或官服。到了清朝，文武百官都穿靴，百姓穿靴并不多见。

中国古代的袜子称为"足衣"或"足袋"，早期用兽皮制成，后来主要用各种布料或丝线制成。在古代，袜子不像现在这样普及，穿袜子是富贵官宦人家的权利，也是一种身份的象征。

previous dynasties and they were used as formal wear or court costume. It lasted to Qing Dynasty (1644-1911) when all officials wore boots while it's rare to see among the ordinary people.

Socks were referred to as "clothes for foot" or "bag for foot". Socks were made of animal skins in early times. Later socks were mainly woven by various cloth or silk thread. Wearing socks was not common in ancient times. Only those who came from wealthy families or official families wore socks, thus wearing socks was a sort of status symbol.

- 云头履

云头履是一种高头鞋履，用布帛为料，鞋首用棕草，高高翘起，形状似翻卷的云朵，故名。

Yun Tou Shoes (cloud-shaped shoes)

Its name comes from the shape of clouds. Yun Tou shoes are made of silk and cotton with raised toe cap.

- 凤头履

凤头履是一种尖头的女鞋，鞋头部分用昂首的凤凰装饰。起初仅用于宫女，后来在民间普及。五代以后，广为流行，清代多用金银片压模成凤头，镶缀在鞋尖。

Crest-shaped shoes

It's a kind of pointed women shoes. There's a head-up phoenix on the toe top for ornament. Initially it was used among maids in the imperial palace, later it was popular among common people. After Five Dynasties (907-960), it widely spread. People in Qing Dynasty (1644-1911) inlayed the toe top with gold and silver plates which were stamped into the shape of crest.

- 宋代穿弓鞋的女子

弓鞋是女子所穿着的弯底鞋,由于鞋底弯曲,外形如弓而得名。宋代以后,因为女子缠足的脚被称为"弓足",所以鞋也被称为"弓鞋"。

A woman with bow-shaped shoes

Bow-shaped shoes are named after the shape of a bow. After Song Dynasty (960-1279), women started to bind their feet which looked like bows, thus the shoes they wore were also referred to as "bow-shaped shoes".

- 双梁鞋

梁,指的是位于鞋头的装饰,通常用皮革为料,裁制成直条,有一梁、二梁、三梁多种。双梁鞋即因有二梁而得名。

Double *Liang* shoes

Liang refers to the ornament on the toe cap. It's made of leather tailoring straight line. There are many types, such as Single *Liang*, Double *Liang* or Triple *Liang*. The name of Double *Liang* shoes is given by the number of *Liang*.

- 草鞋

草鞋是用芒草为料编成的鞋子,通常为贫困者所用。一般男子出门远行也多穿此鞋。

Straw sandals

Straw sandals are woven by Chinese silvergrass. They are usually worn by the poor. Because they are easy to wear, men like wearing them in their trip.

- 高底弓鞋

高底弓鞋是指缠足女子所穿着的高底鞋,流行于明清时期。制为高跟,跟部用木块衬托。

High-heeled bow shoes

High-heeled bow shoes refer to high-heel shoes worn by women who bind their feet. The high heel is supported by wood bricks. It prevailed in Ming and Qing Dynasties.

- 绣花鞋

绣花鞋是指绣有花纹的鞋履，多用于女子，唐代以后广为流行。

Embroidered shoes

There are embroidered patterns on the shoes mostly worn by women. Wearing embroidered shoes became fashionable after Tang Dynasty (618-907).

- 花盆鞋

花盆鞋是清代满族女子所穿着的高底鞋。鞋底的正中央放置木制的高底，外形上宽下窄，形似花盆，因此得名。

Flowerpot-shaped shoes

Flowerpot-shaped shoes are worn by Manchu women in Qing Dynasty. There is a wooden high sole in the center. The shoes look like a flowerpot with wide top and narrow bottom where the name comes from.

- 皮靴

皮靴通常以牛皮制成，帮高至小腿处，帮上开一道豁口，穿时用小皮条连系。战国以后男女均用，流行于西域少数民族地区。

Leather boots

They are made of oxhide. Upper of the boots reaches to crus with an opening, tying it with strips of leather. After Warring States Period, both men and women can wear the boots, prevailing in the ethnic minorities of the western regions.

- 合缝靴

合缝靴是用数块布帛或皮料缝合而成的靴子，唐宋时期较为流行。

Sewn boots

Sewn boots are stitched by several pieces of cloth or leather. They were popular in Tang and Song Dynasties (618-1279).

- 罗袜

罗袜是用纱罗一类的织物为料制成的袜子。因为其质地柔软轻薄，多用于春夏之季，汉代以来广为流行。

Gauze-woven socks

Gauze-woven socks are soft and light so they are mainly used in spring and summer. They have been popular since Han Dynasty (206 B.C.-220A.D.).

- 清代的夹袜

夹袜是用双层布帛为料制作的袜子，一般在寒冷的冬季用来保暖。

Double-layered socks

Double-layered socks are made of two layers of textiles worn in winter to keep warm.

三寸金莲

三寸金莲是指中国古代女人的小脚，源于女子缠足的习俗，符合古代"女子以脚小为美"的审美观念。当时，女子到了一定年龄，双脚要用布帛缠裹起来，最终形成一种特殊形状，世人称之为"三寸金莲"。中国人形容贵重或美好的事物，总喜欢以"金"字修饰。于是，"金莲""三寸金莲"就成了对缠足小脚的美称。双足缠好后，再穿上布帛或绸缎制成的弓鞋，又被称为"小脚鞋"或"尖鞋"。

- 穿弓鞋的清代女子

A woman with bow-shaped shoes in Qing Dynasty

Golden Lotus Feet in Three *Cun* (Lotus Feet)

"Lotus feet" refers to women's little feet. It's from the custom of foot- binding to carter for the aesthetic of "women with little feet are beautiful" in ancient China. At that time, women at certain age had to wrap their feet with cloth and ultimately their feet would form special shapes, known as "lotus feet". When it comes to the precious and valuable thing, Chinese people prefer using the word "golden" to beauty. Thus, literally translated as "golden lotus" or "golden lotus feet in three Cun" becomes the reputation of foot-binding. After binding the feet, women put on the bow-shaped shoes which are made of satin or silk, also known as "pointed shoes" or "little feet shoes".

清代绣花弓鞋
Bow-shaped embroidered shoes of Qing Dynasty

胡服与戎装
Hu Dress and Martial Attire

中国古代的士兵们所穿的军服称为"戎装"。虽然历代戎装式样各不相同，但基本形式都是以身穿甲胄作为防护。中国军服的历史，最早可以追溯到战国时期的"胡服骑射"，从那时起，中国才有了正式的军服。

"*Rong Zhuang*" referred to martial attire worn by ancient Chinese soldiers. Even though martial attire varied in style, the basic form remained having armour for defense. The history of Chinese martial attire could be traced to Warring States Period. The formal martial attire started from "Wearing *Hu* Dress and Shooting on Horse" of Warring States Period (475B.C.–221B.C.).

> "胡服骑射"

公元前307年，赵国的武灵王颁胡服令，推行"胡服骑射"，进行了服饰史上的一次重要变革。

胡服，是指当时"胡人"（西北方的少数民族）的服饰。秦汉以前，中原地区的服装都是宽衣博带，且服装配套极为繁复，不仅穿起来费时，活动起来也极为不便；而胡人服装则是短衣长裤，衣身瘦窄，腰束带，穿靴，非常便于骑射活动。

当时，位于中国西北方的赵国，经常与胡人交战。由于那里的地形多为崎岖的山谷，这使得擅长战车作战的赵国人不占优势。于是，赵国的君王武灵王决定进行军事改革，训练骑兵。既然要发展骑兵，首先就需要进行服装的改革，

> Wearing *Hu* Dress and Shooting on Horseback

In 307 B.C., King Wuling of Zhao State promulgated the decree of *Hu* dress to implement "wearing *Hu* dress and shooting on horse" which was an important reform in the history of costume.

Hu dress referred to the dress worn by the northwestern barbarian tribes in ancient China. In pre-Qin and Han Dynasties, costume of the Central Plains tended to be large and loose. The matching clothes were so complicated that it was not only time-consuming to wear but also inconvenient to dress, while the clothes worn by the northwestern barbarian tribes were in narrow shape tied with the waistband and boots, such costumes were suitable to ride and shoot on horseback.

The state of Zhao, located in

以适应马上作战。

赵武灵王的服饰改革开始了,他吸收了胡人的军服样式,将传统宽衣博带改成窄袖短衣,以便于射箭;将套裤改成有裆的、裤管连为一体的裤子,并用带系缚,这便是裈。裈能保护大腿和臀部肌肉皮肤在骑马时减少摩擦,且裤外不必再加裳。

在鞋帽方面,赵武灵王也进行了改革。胡人的帽子不同于当时中原人的冠,是用于暖额和防风沙

northwestern China often engaged with the northwestern barbarian tribes. Facing the rugged valleys terrain, people of Zhao lost their original advantage in fighting on chariots. Considering their drawback, King Wuling of Zhao State decided to implement the military reform of training cavalry. The priority was given to costume reform so as to meet the requirement of battles on horseback.

The reformed martial attire absorbed the style of *Hu* dress. The reform included changing the traditional loose

- 身穿胡服的胡人
A man of the northwestern barbarian tribes in *Hu* dress

的，由动物皮革制成。于是，赵武灵王下令士兵用黑色的绫绢围戴在头上，后来逐渐发展成为帽。至于鞋，胡人穿的是皮靴，软硬适中，便于跑动；鞋帮很高，一直到膝盖下面，用以护腿，也适于骑马。赵武灵王下令士兵们也穿上这种靴子，并在靴面和靴筒表面还装饰着几十、甚至上百个青铜泡，更显威武夺目。

赵武灵王改革后的服饰逐渐演变改进为后来的盔甲装备。胡服的推广，不仅使赵国屡次赢得胜利，还开创了中国骑兵史上的新纪元。同时，这项服饰改革也弱化了传统服饰的身份等级标示功能，强化了其实用性，对此后的服饰发展产生了重要的影响。

从此以后，"习胡服，求便利"便成了中国服饰变化的总体倾向，汉族人不断吸取少数民族的服饰精华来丰富自己的服饰文化。

costume into short clothes with narrow sleeves for the convenience of shooting arrows, changing the leggings into pants with crotch and pant legs as a whole. Strapped pants were known as Kun. It was used to protect the thigh and buttocks muscle against much friction on horseback. No additional dress was added to pants.

He also carried out a series of reform on shoes and hat. Hu's hat was made of the leather or hide to prevent against sand and to keep forehead warm which was different from the hat of the Central Plains. King Wuling of Zhao State ordered his soldiers to wind the black silk around their heads. Later it evolved into hat. In terms of shoes, *Hu* troops wore leather boots which were neither too hard, nor too soft. The upper of the boots reached below the knees to protect the legs and fit for horse riding. King Wuling ordered his soldiers to wear this style of boots. To show his mighty army, dozens even hundreds of dazzling bronze bubbles were decorated on the upper of the boots.

After a series of reform, China's earliest formal martial attire gradually developed into armor equipment. With spreading *Hu* dress, the state of Zhao not only fought many brilliant battles but also inaugurated a new era in China's cavalvy history. In the meantime, the reform of the dress weakened the symbolic representation of ranks and social status and strengthened the practicability of the costume, which had profound influence in the later development of costume.

From then on, "getting used to wearing *Hu* dress for convenience" has become the general tendency of china's costume innovation. Han nationality has been absorbing the essence of ethnic minorities to enrich its own costume culture.

• 汉代的骑兵
A cavalvy in Han Dynasty

> 甲胄

甲胄是中国古代士兵的主要装备，甲即铠甲，胄即头盔。

早期的甲胄只遮住头、胸等人体的要害部位，制作材料多为野兽的皮、林中的藤木等。其中皮质的护具是由整片皮革制成，披戴在前胸后背，四肢部分不着甲，以免影响活动。

商周时期，根据护体部位的不同，将整片的皮革裁剪成大小不同、形状各异的甲片，然后在甲片上穿孔，用索条编缀成甲。青铜兵器难以刺穿，防护力相当强。西周时期还出现了以青铜甲片编缀铜甲的尝试。

春秋战国是诸侯争霸、群雄割据的时代，从文献记载中可以详细地了解到制作皮甲的复杂程序、工

> Armor

Armor was the main equipment for the ancient Chinese soldiers. *Jia* in Chinese character referred to barde and *Zhou* referred to helmet.

The early armor only covered people's head, chest and important parts of the body. It's made of hide and lianas. The leather parts of the armor were made of a whole piece of leather covering the chest and back. There's no barde on arms and legs so as not to affect moving about.

In Shang and Zhou Dynasties (1600B.C.-256B.C.), a whole piece of leather was cut into plates in different sizes but shapes. Plates were drilled and were connected together to be armor. It's so strong that bronze weapons were hard to pierce. It's a trial for weaving bronze plates with brass plates in western Zhou Dynasty (1046B.C.-771B.C.).

The Spring and Autumn and Warring

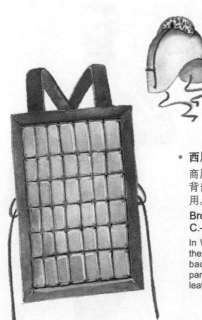

- 西周时的铜甲胄

商周时期，古人把青铜锻打成片，钉缀在胸部、背部，起到类似于盾的保护身体重要部位的作用。铜甲为皮甲的附属物，未能取代皮甲。

Bronze armor of Western Zhou Dynasty (1046B. C.-771B.C.)

In Western Zhou Dynasty, the ancient Chinese forged the bronze into pieces weaving them on the chest and back just like a shield to protect people's important parts of the body. Bronze armor was the attachment of leather armor but it was not replaced by leather armor.

艺以及甲胄的形制、尺寸、结构和各部位的比例。这一时期出现了铁铠，主要用于防护手臂，此后随着铁制兵器的发展，铁铠逐渐代替皮甲成为主要的防护装备。汉代的铁甲在编制工艺上日益精湛，铁甲的锻造技术也不断提高。此时称铁甲为"玄"，即黑色的意思。

魏晋南北朝时期，上好的铠甲用"百炼钢法"锻造。随着重甲骑兵的崛起，适用于骑兵装备的裲裆铠极为盛行，逐渐成为铠甲中的重

States Period (770B.C.-221B.C.) witnessed the feudal princes vying for supremacy and fragmentation of a country by rivaling warlords. In the literature, it recorded the complicated process of leather armor elaborately including its shape, size, structure and the proportion of various parts. At this period of time, the iron armor appeared. It mainly used for protection of arms. With the development of iron weapons, leather armor was gradually replaced by iron armor to become the major protective equipment. In Han Dynasty (206B.C.-

- **秦代士兵的铠甲**

  甲衣的长度前后相等，其下摆一般多呈圆形，周围不另施边缘。肩部甲片的组合与腹部相同，但肩部甲片是上片压下片；为便于活动，腹部甲片是下片压上片。所有甲片上都有甲钉，其数二、三、四不等，肩部、腹部的甲片用连甲带连接。

  Soldier's barde of Qin Dynasty (221B.C.-206B.C.)
  Front length and back length of barde remained the same. The hem was round but there was no edge around it. Plate combination on the shoulder and on the abdomen remained the same. To facilitate moving around, plates on the shoulder were the upper ones holding down the lower ones while the plates on the abdomen were on the contrary. There were nails on each plate. The number of nails varied from one to four. Plates on the abdomen were connected by plate strap.

- **魏晋南北朝时期的裲裆铠**

  裲裆铠主要护卫前胸和后背这两个部位，材料大多采用坚硬的金属或皮革。甲片有长条形和鱼鳞形两种，小型的鱼鳞甲片更便于俯仰活动。为了防止金属甲片磨损肌肤，士兵在穿着裲裆铠时，里面还常衬有厚实的衣衫。

  Liang Dang barde of Wei, Jin, Southern and Northern Dynasties (220-581)

  Liang Dang barde was made of metal and leather used to protect people's chest and back. There were long- strip-shaped and fish-scale-shaped plates. Small fish scale plates were easy for bending. Heavy clothes were worn inside of *Liang Dang* barde to prevent the skin from being abraded by metal plates.

要类型。

　　隋唐时期，裲裆铠仍很盛行。裲裆铠的结构比前代有所进步，甲身由鱼鳞等形状的小甲片编制，长度延伸至腹部，大大增强了腰部以下的防御。唐代还盛行穿明光铠，比较常用的还有绢布甲。绢布甲是用绢布一类纺织品制成的，结构比较轻巧，外形美观，但防御能力较

220A.D.), the process of armor became increasingly sophisticated and armor forging technology was also upgraded. At that time, armor was called *Xuan* meaning black color.

　　The superior armor was forged by "steel smelting method" during Wei, Jin, Southern and Northern Dynasties (220-581). With the rise of heavy-armored cavalry, cavalry was equipped with

- **唐代的绢布甲**

  绢布甲是以绢布为材料制成的铠甲，膝部以上皆能防护。绢布甲对防御远程射击兵器有效，但抵挡不住近战时刀、枪之类冷兵器的劈刺。

  Tough silk armor of Tang Dynasty (618-907)

  Tough silk armor of Tang Dynasty was made of tough silk protecting the body above knees. It could defense against the long-rang shooting but it couldn't withstand the cold steel attack.

- **唐代的明光铠**

  明光铠的前身和背后有圆护，因圆护大多以铜铁等金属制成，并且打磨得极光，战场上在太阳照射下会发出耀眼的"明光"，故名"明光铠"。

  ***Ming Guang* armor of Tang Dynasty (618-907)**

  There were round-shaped protectors on the front and back of the armor. Round protectors were made of copper or iron. In the battlefield, the polished round protectors gave out "dazzling bright" under the sunshine, hence its name.

差，是武将平时服饰或仪仗用的装束。盛唐时期，国力昌盛，四海升平，大部分戎装脱离了实战功用，变成美观华丽的礼仪服饰，不仅铠甲要涂上色，连内衬的战袍也要绣上凶禽猛兽。

宋代的戎装，一种用于实战，一种用于仪仗。用于实战的甲胄防护效果大大加强，据《宋史》记载，整副甲胄共有1825片甲叶，各部件由皮线穿连，总重量约为25公斤。至于仪仗队的士兵服装，多以黄帛为面，布为里，表面绘有绿色的纹样，并以红锦缘边，以青布为下裙。此时还出现了一种既减轻重量又不降低防护能力的轻型铠甲，相传是用一种特殊的蚕茧纸制成的，优点是轻便且防护力较高。

元代铠甲的内层用牛皮制成，外层为铁网甲，甲片相连如鱼鳞，箭不能穿透，制作极为精巧。另外还有各式皮甲、布面甲等。

明代，由于火器的不断更新，厚重铠甲的防护能力相应下降，只有将军全身披甲胄，士兵多着棉甲。虽然棉甲防穿刺型冷兵器的性能不强，但十分轻便，适于野战。

Liang Dang armor which was extremely popular and became the important type of armor.

During Sui and Tang Dynasties (581-907), *Liang Dang* armor was still prevalent. Its structure improved greatly. Armor was woven by small scales-shaped plates with the length extending to the abdomen to enhance the defense of the waist greatly. *Ming Guang* armor was also popular in Ming Dynasty (1368-1644). Tough silk armor which was made of textile was commonly used. It's light and pretty but lack of defense. It's the military commanders' daily costume and it's also used for ceremonial dress. In the prosperous Tang Dynasty (618-907), the nation was in peace and prosperity. Most military uniforms were away from the military function and turned to the beautiful and luxuriant ceremonial costume. The armor was painted the color and even the patterns of ferocious beasts were embroidered on the lining of the coat armor.

There were two types of martial attire in Song Dynasty (960-1279). One was used for combat. The other was used in the ceremony which guards of honor carried flags and weapons. The defensive function for combat was

- 宋代骑兵的瘊子甲

  瘊子甲是一种十分坚精的盔甲，柔薄坚韧，甲片呈青黑色。由于采用冷锻法加工，当甲片冷锻到原来厚度的三分之一以后，其末端留下像筷子头大小的一块，隐隐约约像皮肤上的瘊子，故名。

  ### Chevalier's wart-shape armor of Song Dynasty (960-1279)

  It was a light but strong armor. Plates were in lividity. As a result of cold forging process, when plates were forged above one third of the original thickness, the end of the plates would leave the wart-like piece, which had the similar size to the head of chopstick, hence its name.

greatly enhanced. According to *History of Song Dynasty*, there were 1,825 pieces of plates in total on an entire armor. All components were connected by rubber-covered wire with a total weight about 25kg. Honor guards wore the dress with yellow silk outside, cloth lining, red rim and cyan cloth as skirt. A kind of light armor appeared at that time, but its weight didn't reduce its protective function. According to legend, it's made of a special cocoon paper. The advantage of this armor was light weight and high-performance in protection.

In Yuan Dynasty (1271-1368), the inner layer of armor was made of ox hide. Outer layer was made of wire plates. Plates were connected together just like scales. It's so sophisticated that arrows couldn't penetrate. There were other kinds of plates such as leather plate, cloth plate and so on.

In Ming Dynasty (1368-1644), due to the constant renewal of firearms, the protective capability of heavy armor declined accordingly. Only the general wore the armor. Soldiers wore cotton armor. Although cotton armor was not strong enough to protect against cold steel penetration, it was light and suitable for field operations. It gave better protection

同时能较好地防护火器攻击。

清代军队中的铠甲只在操练、秋阅或仪仗中使用,以示雄武。这一时期是中国古代戎装发展中变化最大的一个时期。一是满族作为统治者对汉族戎装加以改造;二是由于火枪、火炮的普遍运用,导致戎装的变革。清末,中国的水兵、陆军、巡警等服装,已明显具有西欧军服的特征。

against firearms' attack.

In Qing Dynasty (1644-1911), armor was only used in autumn parade, drilling and guard of honor to demonstrate their power and might. This period witnessed the great change in the development of ancient Chinese martial attire. Manchu as the ruling class reformed the martial attire of Han troops. On the other hand, the universal use of guns resulted in changes in martial attire. At the end of the Qing Dynasty (1644-1911), China's navy, army and patrol have their own uniform with distinctive characteristic of Western European military uniform.

- **明代步兵的棉甲**

  棉甲颜色以红、白为主。骑士多穿对襟甲,以便乘马,步兵大多戴一软帽。

  Infantry's cotton armor of Ming Dynasty (1368-1644)

  Cotton armor was in red or white. Chevalier mostly wore cotton armor with opposite front pieces for the convenience of horse riding. Infantry mainly wore a soft hat.

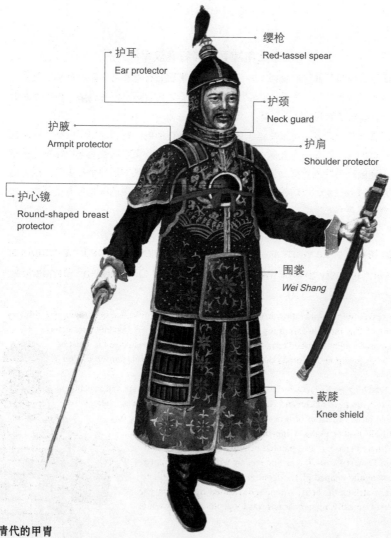

- 护耳 / Ear protector
- 缨枪 / Red-tassel spear
- 护腋 / Armpit protector
- 护颈 / Neck guard
- 护心镜 / Round-shaped breast protector
- 护肩 / Shoulder protector
- 围裳 / Wei Shang
- 蔽膝 / Knee shield

- **清代的甲胄**

  清代铠甲分甲衣和围裳。甲衣肩上装有护肩，护肩下有护腋；另在胸前和背后各佩一块金属的护心镜，镜下前襟的接缝处另佩一块梯形护腹，名叫"前挡"。腰间左侧佩"左挡"，右侧不佩挡，留作佩弓箭和箭囊等用。围裳分为左、右两幅，穿时用带系于腰间。在两幅围裳之间正中处覆有质料相同的虎头蔽膝。

  ### Armor of Qing Dynasty (1644-1911)

  Armor of Qing Dynasty included two parts, *Jia Yi* and *Wei Shang*. There were shoulder protectors on *Jia Yi* and two armpit protectors beneath them. There was a metal round-shaped breast protector on the chest and on the back respectively. Below the round-shaped breast protector was a ladder-shaped protector on the seam of the front part, named "front defense". There was a "left defense" on the left of the waist but none on the right reserving for wearing bow and arrows. *Wei Shang* had left part and right part tied around the waist. Between two parts of *Wei Shang* in the center was covered by the tiger-head knee shields which had the same material as *Wei Shang*.

从秦始皇陵兵马俑看秦汉军服

秦始皇陵兵马俑坑的发掘，对研究秦汉军事服装有着非常重要的意义。至今，共发掘出战车百余乘、陶马600余匹、陶俑近8000件。这支阵容庞大、组织严整的禁卫军，象征了秦朝威武的军队。

观赏这些兵马俑，可以总结出秦汉军服的特征形制。将领的铠甲以整片皮革或织锦材料制成，上面嵌有金属或犀牛皮、野牛皮甲片，高级将领的铠甲除制作精细以外，还绘有美观的纹样。秦代士兵的铠甲由正方形或长方形的甲片编缀而成，铠甲里面要衬以战袍，以防擦伤身体。此外，兵种不同，铠甲的样式也有所不同：步兵的铠甲衣身一般较长，骑兵的铠甲衣身一般较短。

From terra-cotta warrior and horse in Mausoleum of the First Emperor of Qin Dynasty to see the military uniform of Qin and Han Dynasties (221B.C.- 220A.D.)

The excavation of terra-cotta warriors pit has a profound significance in studying the military uniform of Qin and Han Dynasties. Hundreds of war chariots, six hundred terra-cotta horses, nearly eight thousand terra-cotta figures have been excavated so far. Guards with large lineup and rigorous organization symbolize the mighty troop of Qin Dynasty.

Watching terra-cotta warriors can better understand the feature and structure of military uniform of Qin and Han Dynasties. General's armor was made of a whole piece of leather or brocade inlaid with metal, rhinoceros skin or buffalo leather plates. High-ranking officer's armor was not only finely made but also painted beautiful patterns. Soldiers in Qin Dynasty wore the armor woven with square-shaped or rectangular-shaped plates. Coat armor was worn inside to avoid scratching the body. Armor styles varied from infantry to cavalry. Generally, infantry armor was longer than cavalry armor.

- **秦代的石甲**

秦始皇陵兵马俑部分身穿石甲，甲衣上的石片用手工磨制，大多只有0.3厘米厚，是随葬的冥器。

Stone armor of Qin Dynasty (221B.C.-206B.C.)

Some of terra-cotta warriors wore stone armor. Gallets on the stone armor were ground by hand. Most of gallets were only 0.3cm in thickness as burial objects buried with the dead.

婚服与丧服
Wedding Dress and Mourning Dress

在中国传统文化中,"红事"(婚嫁)和"白事"(丧事)都是大事,即人们常说的"红白喜事"。但服饰方面,婚服与丧服的样式和颜色差异却很大。

In traditional Chinese culture, both "red matter" (wedding) and "white matter" (funerals) are great events. In terms of costume, wedding dress and mourning dress have great difference in the style and color.

> 婚服

在中国古代，举行婚礼可谓是人生当中最重要的事情之一，新郎和新娘的婚服自然也是隆重而极具特色的。婚服具有统一的色彩、样式和寓意，用以妆点婚礼，渲染气氛，以及表达美好寓意。

婚服在各个朝代和时期均有所差异，伴随西周礼服的出现，婚服也应运而生。几千年来，中国古代的婚服制式主要有三种，分别是"爵弁玄端——纯衣纁袡""梁冠礼服——钗钿礼衣"和"九品官服——凤冠霞帔"。

西周时，婚礼服饰崇尚端正庄重，与后世婚服区别较大。婚服的颜色要遵循"玄纁制度"，即婚服的颜色为玄黑色和纁红色，大小官员和贵族皇亲都要严格遵守，这种制度一直延续到南北朝。新郎的服饰

> Wedding Dress

In Ancient China, wedding ceremony can be described as the most important event in life. It's natural that bride's and groom's dress are grand and unique. Wedding dress has a unified standard for color, style and meaning for the purpose of adding atmosphere and expressing good wishes.

Wedding dress varied in different dynasties. With the emergence of formal attire in Western Zhou Dynasty (1046B.C.-771B.C.), wedding dress came into being. Three main wedding dress styles appeared in the past several thousand years as shown below.

Wedding dress of Western Zhou Dynasty advocated solemn style which differed greatly from the future wedding dress. Wedding dress followed *Xuan Xun* institution. The color of wedding dress was black and crimson red.

玄纁制度

"玄"即黑中扬赤，象征"天"，古人尊其为天之色彩，较之青、赤、黄、白、黑等五正色尤为尊贵而独居其上。"纁"即赤黄色，象征大地。"玄纁"，即取天地间最高贵的色彩，庆祝一对新人共结连理。

Xuan Xun institution

Xuan namely the color of black faintly red symbolized "heaven". The ancient Chinese named *Xuan* respectfully the color of heaven, which was superior to cyan, red, yellow, white and black. *Xun* referred to the color of red faintly yellow symbolizing the earth. *Xuan Xun* represented the noblest color in the heaven and earth to congratulate the married couple.

当时称为"爵弁"，玄冠、玄端、纁履；新娘的服饰称为"纯衣纁袡"，亦为玄色。

唐代的婚服兼有此前的庄重神圣和后世的热烈喜庆。新郎的服饰为绯红色，头戴梁冠；新娘的服饰为青绿色。晚唐时期，新娘的婚服制式

Officials, nobles and royalties followed this institution until the Southern and Northern Dynasties(420-589). The groom's dress was called "*Jue Bian, Xuan Guan, Xuan Duan, Xun Lv*. The bride's dress was known as *Chun Yi Xun Ran*, namely in black.

Wedding dress of Tang Dynasty

- **唐代命妇的钗钿礼衣**

钗钿礼衣为杂色，由蔽膝、大带、革带、履袜、双佩及小绶花钗等组成，头饰花钿。根据命妇的等级，花钿的数目不等，一品九钿，二品八钿，三品七钿，四品六钿，五品五钿。

Formal dress of *Ming Fu* in Tang Dynasty (618-907)

Formal dress was in variegated color. It consisted of knee shield, sash, leather sash, socks, two ornaments at the waist and flowers. The quality of floral jewelry varied from the ranks of *Ming Fu*. It broke down the first rank having nine floral jewels, the second rank having eight floral jewels, the third rank having seven floral jewels, the fourth rank having six floral jewels and the fifth rank having five floral jewels.

则在宫廷命妇的礼服的基础上发展成"钗钿礼衣"。钗钿礼衣层数繁多，穿着时层层压叠，再在外面套上宽大的上衣；头发上插簪，额间绘花钿。五代以后，繁复的"钗钿礼衣"有所简化。

到了明代，普通百姓在结婚时，男子可穿九品官服，头戴红帽，长衫马褂；女子可穿命妇衣装，身着蟒袍，腰围玉带，凤冠霞帔。

清代有所谓"降男不降女"的规定，因此从清代至民国初年，婚

(618-907) combined the sacred and solemn element in previous dynasties with joyful element in later dynasties. The groom was in crimson and wore *Liang* coronet. The bride's dress was green. In late Tang Dynasty, bride's wedding dress was developed from the costume of women who were given ranks by the emperor (*Ming Fu*) into *Chai Dian Li Yi*. *Chai Dian* dress had many layers and a loose overcoat was over it. Bride wore hairpins and put *Hua Dian* on the forehead. The complex *Chai Tian Li Yi* was simplified after the Five Dynasties (907-960).

In Ming Dynasty (1368-1644), when ordinary people had wedding ceremony, men were allowed to wear the ninth rank of official dress, red hat, gown and mandarin jacket. Women could wear *Ming Fu* dress, coronet, embroidered robe and jade waistband.

In Qing Dynasty (1644-1911), there was a regulation " women wearing the dress of previous dynasty while men are not". Following this regulation, groom wore Qing style costume, that is, cyan robe, black faintly red mandarin jacket over it, winter hat inserted with floral patterns in golden red and red silk over the body. The bride of Han nationality

● **红喜裙**
清代的女子婚服，以大红色绣花，常与大红色或石青色地的绣花女褂配套。

Red Lucky Dress
Wedding dress of Qing Dynasty (1644-1911)was embroidered with red flowers to match red or stone-colored embroidered clothes.

嫁时男子需穿清装，身穿青色长袍，外罩绀色（黑中透红）马褂，头戴暖帽并插赤金色花饰，身披红帛。而汉族女子仍可着明服，身穿红地绣花的袄裙（满族女子着旗袍），外面再"借穿"命妇专用的背心式霞帔，头上簪红花，拜堂时蒙红色盖头。

still wore Ming style dress. She wore red embroidered skirt (Manchu women wore Chi-pao) and "borrowing" *Ming Fu*'s exclusive use vest-style *Xia Pei* over it. Bride wore red flower hairpins. When performing the formal wedding ceremony, bride had a red veil over her head.

- 清代的霞帔
Xia Pei of Qing Dynasty (1644-1911)

- 清代的红盖头
新娘子在结婚当天要以红色盖头蒙面，象征着童贞、年轻、纯洁，婚礼结束后由新郎亲手揭开。
Red veil of Qing Dynasty (1644-1911)
The bride wore the red veil on the day of marriage and it was unveiled by the groom. Red veil represented virginity, juvenscence and purity.

- 清代满族女子的婚服
Manchu woman's wedding dress of Qing Dynasty (1644-1911)

中国红

在中国，红色被认为是吉祥、喜庆的颜色。周朝时，男子就以穿大红衣裳为贵。清朝时，皇帝在天坛祭祀时必须穿红色的朝服。明清时期，不论是皇亲国戚还是平民百姓，结婚时新人们都身穿红衣红衫，这种风俗沿袭至今。过春节时，门上贴的春联、福字和窗花，鞭炮的包装，以及长辈们装压岁钱的红包等都是红色的。

Chinese-Red

In China, red has been considered an auspicious and happy color. Men wearing red clothes in Zhou Dynasty (1046B.C.-256B.C.) represented wealth and honor. In Qing Dynasty (1644-1911), when emperor worshipped at the Temple of Heaven, he shall wear red robe. In Ming and Qing Dynasties (1368-1911), no matter the new couple from royalties or from civilian families wore in red. This custom has been followed so far. In the Spring Festival, you can see the red couplets on the door, red blessing words on the windows, the red package of firecrackers and the red envelopes with lucky money given by the elders.

- 结婚当天披着红盖头的新娘（图片提供：全景正片）
A bride wearing a red veil on the day of marriage

凤冠霞帔

凤冠霞帔本是命妇的规定着装,是权势和地位的象征,普通百姓是不允许穿着的。凤冠即古代贵族妇女所戴的礼冠。霞帔,即宫廷命妇的一种披肩服饰,大领对襟,大袖,左右胯下开叉,绣有云凤花卉,因色彩好似红霞而得名。由于凤冠霞帔上布满了珠宝锦绣,雍容华美,因而逐渐演变成豪门闺秀的婚礼服。再后来,普通百姓结婚时,也可享用"凤冠霞帔",穿一身大红袄裙,外加大红盖头。

Phoenix Coronet and *Xiapei*

Phoenix coronet and *Xia Pei* were the designated dress worn by *Ming Fu* (a woman in ancient China who was given rank by the emperor) which symbolized the power and status. For ordinary people, it's not allowed to wear. Phoenix coronet referred to the ceremonial coronet worn by the ancient aristocratic women. *Xia Pei* was a kind of shawl with opposite front pieces, big collar and slit below each flank of the crotch. It's embroidered with phoenix. The color looked like the setting sun, hence its name *Xia Pei*. Beautiful jewelry was all over the coronet and *Xia Pei*, gracefully and gorgeously, so it gradually evolved into the women's wedding dress of the wealthy and powerful family. Later phoenix coronet, *Xia Pei* and a big red veil were used in ordinary people's wedding.

● 凤冠霞帔(图片提供:全景正片)
Phoenix Coronet and *Xia Pei*

> 丧服

丧服是居丧时所穿服装。中国早在西周时期就已初步形成了丧服制度，针对尊卑、长幼、男女、亲疏不等的各种关系，设计出了质地精粗不同的丧服。两晋南北朝时期，中国的丧服制度正式确立，共分斩衰、齐衰、大功、小功、缌麻等五种服制。

- 斩衰

斩衰在"五服"制度中列位一等。衣和裳分制，衣缘部分用毛边，用最粗的生麻布制成。因为截断的地方外露，故名。凡子、未嫁之女为父母，承重孙为祖父，妻妾为夫，臣为君等服丧皆用此服。服期为三年，除去本年，实际为两周年。

Garb of unhemmed sackcloth

It ranks the first among the five degrees of mourning dress. It's made of the coarsest raw linen cloth with rough edges around the hem. The cut off area is exposed hence its name. Some specific people have to wear it in three years irrespective of that very year, actually two years. They are unmarried girl for her father, grandson for his maternal grandfather, wife and concubines for their husband, ministers for the emperor.

> Mourning Dress

Mourning dress was worn in the funeral. Mourning dress institution was initially set up in Western Zhou Dynasty (1046B.C.-771B.C.). Considering such relations among family members as ranks, genders and degrees of kinship, the mourning dress was designed in different textures. During the Western and Eastern Jin, Southern and Northern Dynasties (265-581) China's mourning dress institution was established formally. It was fallen into five degrees, including garb of unhemmed sackcloth, garb of trim seam, garb of fine linen cloth, mourning dress of the next to last degree and funeral linen clothing of the lightest degree. After Qin and Han Dynasties (221 B.C.-220A.D.), it's been in use with a little change in institution until the early Republic of China.

According to five degrees of mourning dress, the closer the kinship

齐衰

齐衰在"五服"制度中列位二等。衣和裳分制，缘边部分缝缉整齐，有别于斩衰的毛边，由此得名。具体服制及穿着时间视与死者关系亲疏而定。

Garb of trim seam

It ranks the second among the five degrees of mourning dress. The hem has trim seam which is distinguished from garb of unhemmed sackcloth, hence its name. When and who wears it all depends on the kinship with the deceased.

大功

大功在"五服"制度中列位三等，次于齐衰。形制与齐衰相同，但质料不同，是用熟麻布制成，质地较齐衰为细，较小功为粗。大功所用的麻均需剥皮、浸沤、煮练处理，再水洗、碓击，使麻布柔软顺滑。此服期为九个月。

Garb of fine linen cloth

It ranks the third among the five degrees of mourning dress, next to garb of trim seam. It has the same style as garb of trim seam. It uses the fine linen cloth. Its texture is finer than garb of trim seam. Linen cloth is peeled, soaked, boiled to refine, washed again and beaten to make it smooth and soft. Wearing term lasts nine months.

小功

小功在"五服"制度中列位四等，次于大功。用麻布为材料制作而成，质地较大功为细，较缌麻为粗。凡男子为叔伯祖父母、堂伯叔父母、再从兄弟、堂姐妹、外祖父母，女子为丈夫之姑母姐妹及妯娌服丧，均用此服。服期为五个月。

Mourning dress of the next to last degree

It ranks the fourth among the five degrees of mourning dress. It's made of fine linen cloth coarser than funeral lined clothing of the lightest degree. Men wear it for the funeral of their uncles, maternal grandparents and paternal cousins. Women wear it for husband's aunts, cousins or sisters-in-law. Wearing term lasts five months.

缌麻

缌麻在"五服"制度中为重量最轻的一种，用精细的熟麻布制成。凡为本族曾祖父、族祖父母、族父母、族兄弟，以及为外孙、甥、婿、岳父母、舅父等服丧，皆用此服。服期为三个月。

Funeral linen clothing of the lightest degree

It's the lightest degree among all five. It's made of refined linen. It's worn for the funeral of native grandfather, native grandparents, native brother, grandson, nephew, son-in-law, father-in-law and maternal uncle. Wearing term lasts three months.

秦汉以后一直沿用，只是制度稍有变化，直到民国初年。

　　按五服制度规定，血缘关系越近，则服制越重；血缘关系越疏，则服制越轻。在丧葬仪式上，亲属会披麻戴孝，旁人忌穿华丽的衣服。

was, the heavier the mourning dress was and vice versa. In the funeral, family relatives would wear lined clothes and wrap a piece of black cloth around the sleeve. People present avoided wearing fancy costumes.

披麻戴孝

　　在中国古代，长辈去世，子孙们都要身披麻布服，头上戴白，胳膊上裹孝布，以示哀悼，称为"披麻戴孝"。孝服的颜色为白、黑、蓝和绿。儿子、女儿和媳妇的关系最为亲密，因此要穿白色的麻布衣裤。"戴孝"则是在衣袖的上端戴上黑色的孝布，如果死者为男性，则戴在左袖；若为女性，则戴在右袖。

Pi Ma Dai Xiao (wearing linen clothes and a piece of black cloth wrapping around the sleeve to show filial piety)

In ancient China, children and grandchildren wore linen clothes, white cloth on hat and a piece of cloth around the arm to lament the late elder. It's known as *Pi Ma Dai Xiao*. The color of mourning dress was white, black, blue and green. Son, daughter and daughter-in-law were the most intimate with the late elder so they wore white linen coat and pants. *Dai Xiao* referred to a piece of black cloth around the sleeve. If the deceased was male, it's worn on the left sleeve and worn on the right sleeve for deceased female.

出版编辑委员会

主　　任　　王　民

副 主 任　　田海明　林清发

编　　委　　（以姓氏笔划为序）

　　　　　　王　民　毛白鸽　田海明　包云鸠

　　　　　　孙建君　吕　军　吕品田

　　　　　　吴　鹏　林清发　徐　雯　涂卫中

　　　　　　唐元明　韩　进　蒋一谈

纸上博物馆

印
刀石寄情，篆刻有味

老茶具
冲泡时光，品悟人生

兵器
刀枪剑戟，斧钺钩叉

古钱币
铜绿银辉话沧桑

石
石不能言最可人

古铜器
国之重器，青铜文明

服饰
云之衣裳，华夏之服

传统乐器
五音和谐，古韵悠悠

紫砂壶
紫玉金砂，壶中乾坤

传统家具
起居之用，造物之美

少数民族服饰
风情万千，地域美裳

古代佩饰
环佩叮当，钗钿琳琅

文房清供
方寸清雅，书斋淡泊

扇
引秋生手内，藏月入怀中

金银器
奢华之色，器用之极

景泰蓝
紫铜铸胎，金生婉转

衡器
公正立国，权衡天下

盆景
案头山水，方寸自然

陶器
抟土成器，泥火交融

料器
火中吹料，华美天成

漆器
朱墨华美，品位之具

竹木牙角器
镂刻精巧，雅玩清趣

鼻烟壶
不盈一握，万象包罗

瓷
千年窑火，碧瓷青影

茶
一茗一世界

玉
温润有方，石之美者

文化的脉络

吉祥图案
图必有意，意必吉祥

书法
翰墨千年，纸上春秋

笔墨纸砚
清风明月，文房雅玩

唐诗
锦绣华章，半个盛唐

节日
民族记忆，风俗庆典

国画
水墨丹青，落纸云烟

梅兰竹菊
花中四美，君子之德

宋词
浅吟低唱，词以言情

汉字
横竖之间，方正之道

瑞兽祥禽
德至鸟兽，祈福禳灾

古代教育
教之以道，学而致仕

禅
拈花一笑，佛语禅心

古代交通
旁行天下，方制万里

姓氏
一脉相承，炎黄子孙

古代科学
格物致知，天工开物

古代游戏
嬉戏千年，益智悦心

中国结
绳艺千载，情结中国

古代帝王
风流人物，各领风骚

传统美德
立身之本，济世之道

道教
天人合一，贵生济世

神话传说
创世叙说，远古回响

生肖
属相文化，地支纪年

节气
四时和煦，岁时如歌

兵书
兵家智慧，决胜千里

诸子百家
思想交锋，百花齐放

匾额楹联
留墨贵思量，雅韵岁时长

四大名著
传世奇书，文学丰碑

文明的印迹

徽州
一生痴绝处，无梦到徽州

西藏
雪域秘境，心灵净土

佛像
法相庄严，信仰之体

古典建筑
桂殿兰宫，神工天巧

古镇
烟雨千年，山水故园

长城
秦时明月，万里雄关

名山
三山俊秀，五岳奇崛

颐和园
佛香阁暖，昆明水寒

古典建筑装饰
雕梁画栋，绘彩描金

古桥
赵州遗韵，卢沟晓月

名寺
梵林古刹，清凉世界

名塔
乃至童子戏，聚沙为佛塔

民居
阡陌交通，鸡犬相闻

石窟
凿山镌石成佛国

帝王陵寝
地下宫殿，古冢黄昏

名湖
轻烟拂渚，浓淡相宜

牌坊
旌表功德，标榜荣耀

壁画
飞天无影，粉壁乾坤

秦陵与兵马俑
地下雄师，八大奇迹

名泉
竹林清风，洗盏煎茶

大运河
南北动脉，皇朝粮道

丝绸之路
东西走廊，文明纽带

历史名城
文明驿站，王朝印记

北京中轴线
王者之轴，平衡之道

胡同
北京记忆，市井人家

茶马古道
马行万里，茶香千年

长江黄河
文明之源，华夏之根

故宫
盛世屋脊，紫禁皇城

秦砖汉瓦
秦汉气象，土木之工

历史活化石

园林
山池之美，宛若天成

京剧
生旦净丑，唱念做打

旗袍
优雅风韵，花样年华

木文化
盛木为怀，和木而生

太极
阴阳辩证，无极而生

剪纸
妙剪生花，大千世界

酒
开君一壶酒，细酌对春风

中华美食
烟火人间，味道中国

传统手工艺
镂尘吹影，匠心传世

年画
新桃旧符，迎福纳祥

武术
文以评心，武以观德

民间玩具
泥木之艺，奇趣之具

染织
草木之色，纵横之美

刺绣
针绕指尖，线舞布上

中国色彩
五色人生，多彩中国

雕刻
巧匠神技，托物寄情

中医
悬壶济世，妙手春风

茶艺
壶中真趣，廉美和敬

面具
原始面孔，沟通天地

婚俗
婚姻之道，嫁娶之礼

皮影
隔纸说话，灯影传情

泥塑
传世绝活，妙手出神

面塑
诞生于餐桌的艺术

风筝
好风凭借力，送我上青云

灯彩
花灯如昼，溢彩流光

木偶
笑尔胸中无一物，本来朽木制为身

杂技
惊险绝绝，艺动人心

棋艺
风雅手谈，桌上厮杀

民间戏曲
好戏连台，乡土有味